LABORATORY TECHNIQUES
IN
FOOD ANALYSIS

CW01499109

D. PEARSON
DSc, MChemA, FRIC, FIFST

National College of Food Technology,
University of Reading, Weybridge, Surrey

LONDON   BUTTERWORTHS

THE BUTTERWORTH GROUP

ENGLAND
Butterworth & Co (Publishers) Ltd
London: 88 Kingsway, WC2B 6AB

AUSTRALIA
Butterworths (Pty) Ltd
Sydney: 586 Pacific Highway, NSW 2067
Melbourne: 343 Little Collins Street, 3000
Brisbane: 240 Queen Street, 4000

CANADA
Butterworth & Co (Canada) Ltd
Toronto: 14 Curity Avenue, 374

NEW ZEALAND
Butterworths of New Zealand Ltd
Wellington: 26–28 Waring Taylor Street, 1

SOUTH AFRICA
Butterworth & Co (South Africa) (Pty) Ltd
Durban: 152–154 Gale Street

First published 1973

© Butterworth & Co. (Publishers) Ltd., 1973

ISBN 0 408 70424 1

Filmset by Keyspools Ltd, Golborne, Lancs.
Printed by C. Tinling & Co. Ltd, London and Prescot

WITHDRAWN

Laboratory Techniques Series

LABORATORY TECHNIQUES
IN
FOOD ANALYSIS

PREFACE

The principal aim of this book is to assist the technician carrying out the routine analysis of foodstuffs. In so doing, it is also bound to assist students who are required to carry out practical work in courses on food science and food technology, which require a study of food analysis.

When the author was originally approached by the publishers to write a book on the chemical techniques used in food analysis, it was necessary to decide on the respective weight to be given to the many possible aspects and approaches that might be incorporated. Bearing in mind other works related to the subject that have been published since this book was originally envisaged, together with the aim of producing a textbook that is suitable for use by the food technician, it was decided to cover in the introductory chapter some of the general principles of quality control in the food industry. The early chapters give details of methods that are commonly used for the estimation of the basic components of foods, selected additives and trace metals, and subsequent chapters aim to show how these and other techniques could be applied to some of the more important food products and manufacturing processes. Whilst within the size of this book it is obviously not possible to mention all foods, the author feels that judicious application of the methods described, particularly those in the earlier chapters, should enable the average reader to examine most of the materials that are likely to be encountered.

With certain exceptions, methods that require specialised apparatus associated only with a narrow area of the food industry have been excluded as far as detailed treatment is concerned, although appropriate references are given at the end of each chapter. Some routine methods for the examination of water supplies are given in the last 'Miscellaneous' chapter, and the Appendices include some notes on food legislation, which appears to be of increasing concern to most food analysts.

For basic chemical methods, the reader should refer to the

companion volume *Laboratory Techniques in Chemistry and Biochemistry* (Butterworths, 2 edn 1972) by P. Diamond and R. Denman. Instrumental techniques are covered in Vogel's *A Textbook of Quantitative Inorganic Analysis* (Longmans, 1961) and in Y. Pomeranz and C. E. Meloan's *Food Analysis: Theory and Practice* (Avi Publishing Co., Westport, Conn., 1971).

Finally, particularly as this is a first edition, the author would be most grateful if readers would inform him of any errors or omissions that they notice in order to assist in the preparation of any subsequent edition.

National College of Food Technology,   D. PEARSON
University of Reading,
Weybridge, Surrey, 1972

ACKNOWLEDGEMENTS

The author would like to express his most sincere thanks to Mr G. R. B. Gardiner for assistance given at all stages in the preparation of this volume. Assistance in proof reading was also most willingly given by Mr S. Landsman. Thanks are also due to Mrs M. Fraser who commented from the technician's viewpoint and prepared most of the illustrations and to Mrs B. A. Shore who converted many of the virtually unreadable manuscripts to immaculate typescript.

CONTENTS

ABBREVIATIONS   xi

1 INTRODUCTION   1
Some basic principles of quality control

2 GENERAL METHODS—BASIC CONSTITUENTS   27
Sampling—Moisture—Equilibrium Relative Humidity—
Fat and Volatile Oil—Fibre—Protein—Ash—Sugars—
Acidity—pH Value—Alcohol

3 FOOD ADDITIVES   78
Preservatives—Antioxidants—Colouring Matters

4 TRACE ELEMENTS   97

5 OIL VALUES AND RANCIDITY   119

6 DAIRY PRODUCTS   131
Milk—Cream—Evaporated and Condensed Milk—Milk
Powder—Butter—Margarine—Cheese—Ice Cream

7 FLESH FOODS—MEAT AND FISH   166
Assessment of Freshness—Assessment of Meat Content—
Brines—Cured Meat—Smoked Fish

8 FLOUR, BAKING POWDER AND SELF-RAISING FLOUR   213

9 SUGAR AND FRUIT PRODUCTS   237
Sugar—Jam—Canned Fruits—Pickles and Sauces—
Vinegar—Tomato Purée—Fruit Juices

10 MISCELLANEOUS   270
Hardness of Water—Free Chlorine—Alcohol Insoluble

CONTENTS

Solids—Caffeine—Assessment of Cocoa Content—
Assessment of Egg Content—Solubility of Dried Egg—
Creatine and Creatinine in Meat Extract—Jelly Strength
of Gelatine—Pesticides

APPENDIX I—Notes on Food Legislation   286

APPENDIX II—Notes on Spectrophotometry   290

APPENDIX III—Factors for Food Energy Calculations   294

APPENDIX IV—Control of Carcinogenic Substances   295

APPENDIX V—Preparation of Indicators   296

APPENDIX VI—Factors for Solutions in Volumetric Analysis   298

APPENDIX VII—The SI System of Units   300

ATOMIC WEIGHTS OF THE ELEMENTS   303

LOGARITHMS   305

INDEX   309

ABBREVIATIONS

Alcohol	Purified industrial methylated spirit
AMC	Analytical Methods Committee
AOAC	Official Methods of Analysis of the Association of Official Analytical Chemists
APA	Association of Public Analysts
b.p.	Boiling point
BP	British Pharmacopoeia
BS	British Standard
FACC	Food Additives and Contaminants Committee
FF	Fat-free material
FFA	Free fatty acids
f.p.	Freezing point
FSC	Food Standards Committee
ICUMSA	International Commission for Uniform Methods of Sugar Analysis
LAJAC	Local Authorities' Joint Advisory Committee on Food Standards
MRC	Medical Research Council
MSNF	Milk solids-not-fat
NFS	Non-fatty solids
OP	Over Proof
ppm	Parts per million
PS	Proof spirit
SAC	Society for Analytical Chemistry
SI	Statutory Instrument
SNF	Solids-not-fat
SPA	Society of Public Analysts & Other Analytical Chemists (now the SAC)
S.R. & O's	Statutory Rules and Orders
TMA	Trimethylamine
TVN	Total volatile nitrogen
UHT	Ultra heat treated (milk)

1

INTRODUCTION—SOME BASIC PRINCIPLES OF QUALITY CONTROL

INTRODUCTION

Modern civilisation demands the production of a wide variety of foods in large quantities. To supply the modern retail outlets, increasing attention has been given to the development of relatively small pre-packed units of foods, which have usually been submitted to one or more of the various methods of preservation available including drying, chilling, freezing, heat processing or the addition of chemicals. Also, in order to make products more acceptable to the consumer, other additives are sometimes incorporated into the product, such as artificial colours, emulsifiers and stabilisers. As costing cannot be ignored however, artificial sweeteners are used to replace some of the sugar in soft drinks and butter may similarly be replaced by other fats in flour confectionery. Most foods are therefore now produced so that they consistently have the same effect on the senses in each unit pack. Additionally, weights have to be closely controlled. Consequently, the development of products must include a large number of factors such as composition of raw materials, processing methods, type of plant, compliance with specifications and legal standards, the necessary control methods at the various stages of production, type of package, weight tolerances, storage characteristics and the all-important costing of materials, fuel, services, labour, storage and distribution.

The term 'Quality Control', however, means something different according to the particular organisation. In some cases, quality control refers to a few on-the-spot tests carried out on the factory floor or adjacent to it. In the manufacture of certain products the main techniques involved consist of visual inspections at various stages. There may, however, be a block of laboratories that deal not only with immediate control matters as such, but also with longer term activities associated with production control in general, such as the development of new products and long-term pure and applied research. In these days of tighter budgeting, quality control departments are increasingly concerned with schedules, efficiency of plant and personnel and the statistical evaluation of numerous measurements on the short- and long-term basis. This volume concentrates mainly on the activities revolving around the activity of the chief chemist.

The organisation of quality control is basically associated with the following sequence of operations:

Agricultural control.

Examination of raw materials, most of which have already been partially processed and submitted to some form of standardisation.

Formulation and examination of trial batches.

Process control, which includes examination of 'intermediates' and checking of recording charts on equipment and hygiene inspection.

Product examination, including weight control and compliance with specification and legal requirements.

Storage controls in the warehouse and at the retailer's premises.

Apart from these routine activities the chemist also becomes involved in development work, the preparation of specifications and procedures, the evaluation of the comparative quality of competitors' products, the training of personnel and trouble-shooting. The chief chemist usually acts as adviser on legislative matters and in view of the numerous statutory controls involved may sometimes have to 'tone down' some of the Advertising Department's attempts to use certain immodest types of claims. In view of the importance of integration with other departments, it is most necessary for quality controllers to develop the ability to explain matters in terms that are understandable to associates who would otherwise be blinded by scientific jargon. Simplification and tact in reporting is an art which must be acquired by all those concerned.

THE TERMS 'QUALITY' AND 'QUALITY CONTROL'

Quality can best be considered as the product of various attributes, including composition and nutritional value, purity, appearance, taste, odour and consistency. It should be thought of as their product rather than their sum, because if any one of these attributes is low the general quality will be low. We might produce a food containing every necessary nutrient for health, but if it had an unattractive taste it would be difficult to sell. Commercially, however, variations from the norm are slight and quality is assessed by the summation of attributes.

Quality control, in spite of variability in raw material due to its being of biological origin, means a system which standardises the materials used, the process itself and, in turn, the product. Adjustments to the process should be made during manufacture. This requires the development of suitable rapid techniques of testing even though they may be indirect physical or empirical test methods.

The purpose of quality control is the manufacture of products that are standardised within appropriately selected commercial limits. Although this may ensure that overweight and very high quality goods are not sent out, it equally sees that the customer does not receive sub-standard products.

THE HISTORICAL BACKGROUND TO QUALITY CONTROL IN THE U.K.

If we look back to the middle of the last century, we find that the Lancet Sanitary Commission and Government Select Committees were damning the manufacturers for various frauds and malpractices[1]. In some instances it is apparent that the offences were connected more with ignorance than with deliberate deceit. On the whole, however, it would appear that at that time manufacturers and the authorities were far from working together in a friendly manner. Also, when the first manufacturers set up their own laboratories, the main purpose seems to have been to ensure that products complied with statutory requirements.

It was during the period between the wars that most food manufacturers set up scientific organisations and most firms eventually required their scientific personnel to assist positively in various aspects of the production programme. The first chemists who entered the food factories, however, had first to obtain the co-operation of the experienced personnel on the floor (a problem not unknown today either). They found what Bassett[2] called a rule-of-thumb regime and the scientist had to study the changes occurring in the processes, put recipes on to paper and devise suitable control tests in order to standardise procedures.

Limits had to be worked out for each material, intermediate and product, and storage characteristics were studied in relation to known faults that commonly occurred. Hughes[3] aptly summarised what occurred as adding the 'know-why' to the 'know-how'. As time went on, many firms set up, in addition to control systems, departments to deal with product development and others to pursue longer term research programmes. This required the study of a variety of sciences, which the chemists appointed had often either forgotten or had to learn anew. Gradually more sophisticated techniques for the examination of materials were developed and statistical methods for interpreting results were increasingly used. Sampling procedures, the specification of buyer's and vendor's risk and the visual representation of variations between pre-arranged limits on quality control charts had to be studied (and applied if possible) for each product.

THE MAIN STAGES IN QUALITY CONTROL SYSTEMS

AGRICULTURAL CONTROL

The quality control of some foods begins at the farm. For instance, the best blends of varieties of wheat for bread-making and the most suitable fruits to use for jam-making, canning and freezing are selected and inserted in specifications. Some previously suitable varieties may have become prone to pick up a disease and new types may have to be sought. Also, in vegetable canning a contract crop is grown with the factory either supplying the seed or specifying the necessary variety. With fruits and vegetables, the respiratory activity falls off slightly after picking, then rises steadily to the ripeness stage (the climacteric) and falls away again when over-ripe[4]. Concomitant changes occur in the acid and pectin contents, and with potatoes, apples and bananas the carbohydrate changes markedly such that there is inter-conversion of starch and sugars. The degree, rapidity and direction of the change is affected by the temperature and composition of the storage atmosphere. With fruit the all-important colour also changes and when considered in relation to softness this enables the fruit to be picked just prior to maturity. In jam manufacture, the boiling encourages the conversion of the protopectin to the soluble, more effective pectin to give the best set[5]. Also, if fully ripe fruit is used in canning and bottling, the heat processing tends to soften the fruit and breakdown occurs in the product. Slightly harder under-ripe fruit, on the other hand, softens slightly in the process and a firmer, more desirable product results.

Research into breed selection and feeding gives the factory information on the rapid production of animals and poultry with suitable characteristics. With pigs in particular, the factory demands an animal that will supply a high lean-to-fat ratio. Specifications therefore lay down the breed of pig together with a limit on the weight range. In turn, the farmer seeks economical production by employing appropriate methods of feeding.

RAW MATERIALS

The term 'raw materials' covers all items delivered to the manufacturer, whether used directly in food production or not. Apart from food ingredients and water, therefore, the quality control laboratory examines, for instance, detergents, containers and even packaging materials, fuel and lubricating oil, paint and sometimes building materials and cloth. If the material is to come from new sources of supply, a small *buying sample* is examined to ascertain that it has the desired properties for the intended product. Then the first delivery ordered and every subsequent one should correspond with that sample, unless the supplier gives a statement to the contrary. With raw fruits and vegetables, a certain percentage of defective units, worked out on a statistical basis, is laid down in specifications[6]. The delivered materials are checked to see they are 'up to sample' and in extreme cases the receiver can seek protection under the Sale of Goods Act. More generally, a decision is made for every raw material received as to whether it is to be accepted, rejected, held pending correspondence and discussions, blended with other material or used for a different product. It should be borne in mind that such decisions (as with systems covering the sampling frequency) will depend to some extent on the good name (or otherwise) of the supplier in relation to past experience.

Some aspects of the factory chemist's work are similar to the work carried out in the laboratories of the local authority. Thus raw materials are checked for identity and compliance with legal standards as to minimum quality, the absence of adulterants and limits of impurities such as trace elements and pesticide residues. Specifications, however, may require numerous other regular checks in order to standardise size, colour, clarity, flavour, particle size and aroma, together with other factors that affect the palatability and organoleptic acceptability. A good, although extreme, example of this is the examination of gelatine, which is an expensive processing aid that differs considerably in properties and is submitted to a rather intensive examination in the factory. The comparison is brought out in *Table 1.1*. Such factors as clarity, viscosity, jelly strength and prevention of crystal formation are most important in ensuring that products reach the consumer in the form intended when the original recipe was devised. The necessity of quantitatively controlling the jelly strength can be readily seen in the case of meat pies, for which a weak-setting, runny gel will readily encourage the growth of micro-organisms.

Table 1.1. COMPARISON OF WORK CARRIED OUT ON GELATINE IN DIFFERENT LABORATORIES

Examination for	By Public Analyst*	Factory laboratory	Remarks
Moisture	Seldom	?	Of less importance than usual
pH	No	Yes	Bound up with gelling properties
Jelly strength	Simple statutory test	Quantitative determination	Also effect of pH and heat considered
Viscosity	No	Yes	
Colour	No	Yes	
Clarity	No	Yes	
Odour	Yes?	Yes	
Flavour	Yes?	Yes	
Sediment	Yes?	Yes	
Suitability for product	No	Yes	⎫ gelling-jellies, whipping-marshmallows, stabilising, ice-cream ⎭
Ash	Yes	Yes	⎫
Arsenic	Yes	Yes	
Copper	Yes	Yes	
Lead	Yes	Yes	Statutory limits
Zinc	Yes	Yes	
Sulphur dioxide	Yes	Yes	⎭
Iron	No	Yes	Affects colour
Calcium	No	Yes	Indicates type and whether crystals are formed with tartaric acid
Microbiological examination	?	Yes	
Tests on trial batch of product	—	Yes	

* This refers to examination for 'official' work. Many consultant public analysts carry out analyses for industry and then use tests similar to those applied by the factory chemist.

Other raw materials of interest are colours, which often contain blends of up to five or six different dyes. Apart from identity and purity, it is necessary to control the tinctorial power in view of the importance of standardising the appearance of the product. It is usual, therefore, to prepare from the first delivery a standard graph that relates the optical density at the wavelength of maximum absorption against concentration. Then solutions of subsequent deliveries are prepared, and after comparing the absorption against the standard the tinctorial power can be expressed as a 'percentage of the standard colour' (from the first delivery; see page 247).

When oils and fats are received, the identity, purity and the freshness are checked. If an oil is to be stored, however, its liability to future deterioration can be assessed. This is effected by using accelerated storage tests, which cause a rapid increase in the rate of rancidity formation. In the commoner methods, the peroxide value is measured whilst the fat is aerated at an elevated temperature. In this way the course of rancidity in future weeks can be approximately assessed in a matter of hours.

Flour is blended from wheat obtained from various countries, and different physical characteristics are demanded according to the product. Also, after receipt, the moisture content of the grain may have to be adjusted to make it more suitable for milling. When received at the bakery, the flour is checked for both composition and comparative physical properties. The colour can be checked by visual comparison against a type sample of the expected extraction rate, or photoelectrically by means of the Kent–Jones and Martin Flour Grader. Various methods are also available for predicting the ability of the flour to produce a dough with suitable characteristics. Thus gas production is measured from the maltose figure. Also, various special instruments are available for assessing water absorption, resistance of the dough to mixing (the farinograph) and resistance to stretching (extensograph).

One factor which makes standardisation of the ingredients (and consequently the products) most difficult in food manufacture is the wide variation in composition due to the materials being of mainly biological origin. Such variations are especially apparent with raw flesh, fruits and vegetables. Thus, although the meat content of products is assessed by taking into account the nitrogen figure for the flesh used, the mean factors to be used have to be deduced from a wide range of values obtained from different cuts, etc. The Society for Analytical Chemistry has found it necessary to publish for meat, poultry and fish, agreed compromise N/FF factors, so that disputes on the interpretation of analytical results are less likely. The even wider variation in the composition of fruits causes additional difficulties in assessing the amount used in preserves and the proportion of juice present in squashes.

Nowadays, however, numerous materials as received in the factory have already been submitted to a fair degree of processing and standardisation, e.g., dried and frozen foods, purées, extracts, purified chemicals, starches and oils. Apart from the advantages of controlled composition and properties, such manufactured ingredients are usually cleaner and purer than the natural materials from which they are derived. In particular, crude herbs and spices tend to show widely differing compositions and require storing as the whole spice, because of the more rapid loss of volatile oils after grinding. They also frequently have a high bacteriological count. On the other hand, solid extracts dried on a carbohydrate or salt base, which are ready for use, standardised in composition and virtually sterile, are likely to produce a more consistent controlled product with better keeping qualities[7]. Perhaps one of the main changes in the manufacturing system which has affected the chemist is delivery in bulk rather than in smaller individual units. Free-flowing powders are delivered pneumatically direct from the vehicle into the premises, and syrups and oils are pumped in by pipeline ready for immediate metering into the process itself. Such methods have an effect on the sampling procedure and some tests may be applied at the point of delivery.

TRIAL BATCHES

Trial batches of the product are prepared from certain raw materials, to check whether any recipe modification is required, e.g., trial bakes are made from flour, jellies from gelatine, and the new season's fruit is canned to check general suitability and syrup concentration. During processing the added syrup is diluted by the juice in the fruit and the strength may require adjustment to give constant sweetness on cut-out. For jam, a trial batch of the fruit shows if the processing method gives a product of correct quality and ability to withstand storage. Checks on the pH, soluble solids and reducing sugars are therefore made and adjustments effected where necessary.

The correct metal to use for the equipment is also studied. Any food that contains acid, alcohol or salt is liable to corrode metals. The liability to corrosion can be checked before ordering equipment by maintaining samples of the metal in the food at an elevated temperature. The copper used for new processing equipment has now largely been replaced by other metals such as aluminium, nickel and selected stainless steels. Ordinary stainless steel, which contains chromium and nickel together with small amounts of carbon, silicon and manganese, is not suitable for foods that contain salt or vinegar. As plant may be required for the preparation of a variety of foods during the various seasons of the year, such grades would normally be selected for factory use.

PROCESS CONTROL

Having checked that the raw materials are of a suitable quality and commercially pure, it is necessary to consider whether it is advisable to carry out checks during the process to enable certain modifications to be made. These should have been allowed for when the formulation and processing method were worked out originally. Important considerations here are the best mixing technique to use to attain homogeneity, the processing times and temperatures and the relevant tools, instruments, gauges, meters and charts to check that the procedure is standardised.

Rapid feed-back (feed-forward if possible) of information to the factory floor is all-important in process control, so lengthy methods of testing are inappropriate. During the process, therefore, on-the-spot tests are devised when possible to measure (often indirectly) composition and physical properties. For instance, Brix hydrometers measure the sugar content of syrups rapidly and enable quick adjustments to be made. Similarly, salinometers (or brineometers) give the salt concentration in brines as a percentage of saturation. Automatic gravity control is practised in some processes.

Numerous intermediates are controlled to a definite soluble solids content by using total-reflection refractometers located close to the process, e.g., preserves, jellies, purées, fruit pie fillings. Micro-organisms are liable to grow in such materials unless the finished product contains at least 65% of sugar. In some instances, however, the soluble solids content at which an intermediate is controlled may be lower than this value if water is removed during subsequent heat processing. Such control limits at the various stages of the process are calculated during the development of the product.

In the manufacture of canned and frozen vegetables, the adequacy of blanching is assessed by confirming that the destruction of enzymes such as peroxidase is complete by using guaiacol–hydrogen peroxide solutions.

Viscosity control is important with many foods, e.g., sauces, creams, gelatine solutions and chocolate couverture. Food materials are non-Newtonian (the viscosity alters with changes in the rate of shear) so the reading obtained differs according to the manner in which the measurement is made. Control limits can be worked out by using simple 'comparative' viscometers, e.g., the time for (a) the material to flow out of a funnel (or between fixed points of it) or (b) a metal ball to fall a standard distance through the material.

Consistency or jelly strength is commonly measured on gelatine and pectin gels, which have been set under standardised conditions[8-10]. The gel strength of gelatine is compared with that obtained

9

from the first delivery as represented on the 'standard' graph. The quality of plastic-type products such as apple and tomato purée and sauces can be assessed by using consistometers, which measure the degree of spread or flow of the material under standardised conditions.

Quantabs[11] are small plastic strips that contain an impregnated capillary element, which give a rapid measure of composition (*Figure 1.1*). After tearing off the end-tab, the strip is immersed in

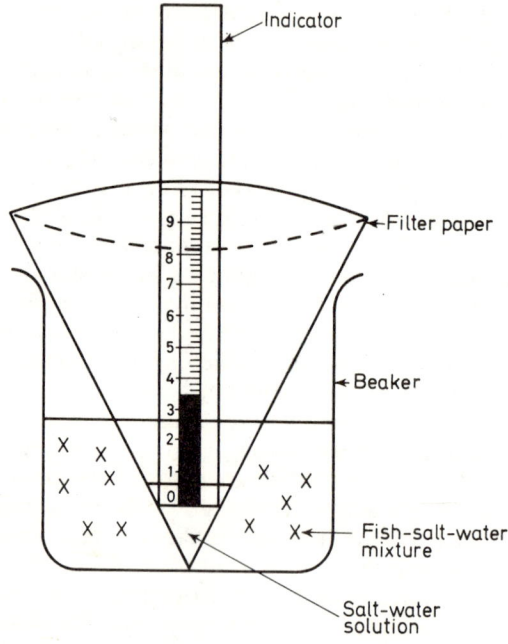

Figure 1.1. Diagram showing degree of reaction in a Quantab plastic strip due to salt in solution. The empirical scale is calibrated against solutions of known concentration of sodium chloride. (From Greig and Seagran, *Comml. Fish Rev.*, **27**, No. 12, 19 (1965), Fig. 1)

the food solution and the concentration is assessed from the height of the coloured column. Quantabs can be used mainly for measuring salt, acidity and alkalinity.

There has been increasing interest in recent years in more automated analytical measurements, and in particular the application of the Technicon AutoAnalyzer in on-stream control[12]. The need for the replacement of manual methods by automated techniques stemmed principally from the shortage of technicians and the consequent uneconomic use of qualified staff. In the Technicon system (*Figure 1.2*), the samples are usually placed in small plastic beakers in the circular sampling plate. Each sample meets the tubes that contain the appropriate reagents at the proportioning pump and the liquids are thus mixed before being passed through a semipermeable membrane which replaces normal filtration. The actual reaction takes place in a thermostatically controlled water-bath. The reaction is not necessarily completed, but every sample in the stream receives the same treatment and, in the calibration, standards are treated in a similar manner. The next stage in the line is some form of electrical measurement—usually a colorimeter or flame photometer and the degree of reaction of each is shown as a peak on an industrial-type recorder. The appropriate reagent concentrations and reaction conditions have to be worked out with considerable care before applying the AutoAnalyzer to the process and regular checks must be made to ensure that the equipment has not gone 'out of adjustment'. The system can be used, for example, in the determination of sugars, ammonia, vitamin C and numerous trace elements. Although a rotating Kjeldahl digester can be incorporated into the system for solids in suspension, the most common applications have been to liquids. Consequently the section of the food industry which can make most use of the technique is the brewing industry[13]. The application of on-stream control may therefore be limited according to the availability of a suitable sampling technique. In general, however, systems are available in which free-flowing powders can be drawn from the stream either continuously or at timed intervals.

In view of the obvious difficulties in devising rapid microbiological methods as such, chemical tests have been devised for indirectly assessing that pathogenic organisms have been destroyed. Thus with milk the destruction of TB is approximately concomitant with that of phosphatase, so the confirmation of the absence of the enzyme by a chemical test indicates that the heat treatment was adequate. Similarly, correctly processed liquid egg should be free from salmonella, as indicated by the destruction of α-amylase. Also, the statutory turbidity test for sterilised milk confirms that higher temperatures have been used, as shown by albumin denaturation. Bacteriological counts are carried out regularly, however, particularly when perishable foods are being manufactured, and the results can be tabulated or recorded on control charts so that changes can be readily seen on a week-to-week basis. A study of such trends gives a general indication of whether or not hygienic standards are being adequately maintained. Daily inspections by qualified personnel play an important part in the control system. The plant should not only appear to be clean, but should be of such a

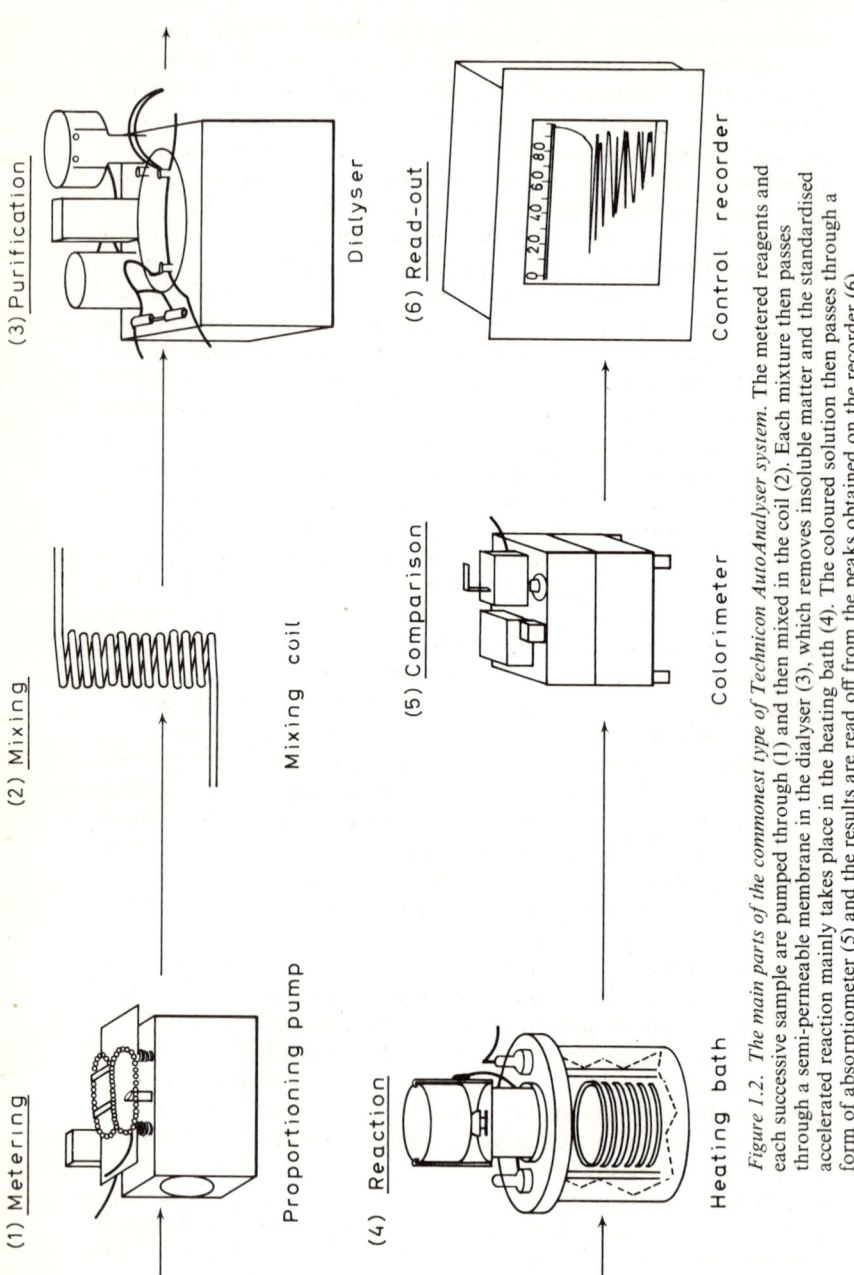

(1) Metering   (2) Mixing   (3) Purification

Proportioning pump   Mixing coil   Dialyser

(4) Reaction   (5) Comparison   (6) Read-out

Heating bath   Colorimeter   Control recorder

Figure 1.2. The main parts of the commonest type of Technicon AutoAnalyser system. The metered reagents and each successive sample are pumped through (1) and then mixed in the coil (2). Each mixture then passes through a semi-permeable membrane in the dialyser (3), which removes insoluble matter and the standardised accelerated reaction mainly takes place in the heating bath (4). The coloured solution then passes through a form of absorptiometer (5) and the results are read off from the peaks obtained on the recorder (6)

12

type that it can be readily cleaned. Plant should therefore be designed to ensure the absence of inaccessible crevices that encourage attack by micro-organisms and insects. Apart from dirt in general, however, regular inspections should be made, to look for (preferably to try to foresee) possible contamination of the products by plasticisers and grease from machinery and belts, loose plaster from ceilings and walls, broken glass from equipment, and so on. There are, however, various types of plant available that enable the manufacturer to reject, or check the presence of, certain forms of extraneous material. Automatic grading plant is used for removing lower quality materials and unwanted matter. Discoloured beans, almonds and rice are deflected electrostatically and peas are separated by flotation. Insects and mites and their eggs in powders such as cereal flours are killed by impact due to passage through special centrifugal machinery. There are also various detection systems in which magnets show up the presence of ferrous metals, electronic devices that detect non-ferrous metals and x-rays that warn of the presence of metals, rubber, stone and possibly glass. Such devices can be adjusted for varying sensitivity and the detector has to be linked with a warning, rejection or process-stopping system. Inspection devices are also available for detecting extraneous matter in glass containers. It should be added that none of these devices represents a 100% safeguard and that they are fairly expensive to instal.

Also, the factory lays down rules for personnel, such as the following:

(a) Containers closed with nails or wire should be opened in rooms away from the process.

(b) Extraneous matter is removed by magnetic and electrostatic devices or by sieving.

(c) Machinery should be regularly examined for loose screws, nuts, bolts, washers and plating.

(d) No glass articles are allowed on the factory floor.

(e) After maintenance, all loose material must be taken away, e.g., wire, metal from drilling, solder and string.

(f) Overalls should have tie-on tapes instead of buttons, and no pockets.

(g) Women should be strongly discouraged from wearing clip-on jewellery.

(h) Protective headgear is provided to prevent hair-pins and hair from falling into the food.

Such instructions have to be supplemented by some form of educational programme or publicity, as more important than rules is the conditioning of personnel into the idea of foreseeing possible troubles.

PRODUCT EXAMINATION

Examination of the product is not truly control, because the product stage is too late to make advisable modifications. Nevertheless, product examination should confirm that the desired results (decided upon when the recipe was formulated) have been achieved. For example, although ingredient purity has been ensured, there may have been contamination pick-up from plant during the process. Also, the ability to withstand storage can be confirmed only on the product itself. Therefore representative samples are drawn from coded batches or streams to confirm that the product complies with the specification and any legal or recognised limits. In addition, there is a considerable amount of psychological value in taking samples in full view of the factory personnel (even though the samples may not be examined!).

A first consideration is weight. At the end of the production line, packets are weighed individually or go to check weighers, which can be set to reject units at any selected level[14]. The weights should in all instances fall between the limits on the appropriate quality control charts.

In the laboratory any individual samples received are also checked for net weight or volume or number. Correctness of composition and properties can be ascertained with a reasonable degree of safety by using relatively simple methods such as measurement of colour and determination of water, fat, pH, salt and soluble solids. Any samples that reveal abnormalities from such tests must be examined further, however, possibly by a more elaborate method. Products that are being made to a quality which is close to minimum statutory or recommended requirements (e.g., certain meat and dairy products) should obviously be submitted routinely to a more frequent and elaborate testing scheme.

Many tests applied to the product are related to organoleptic acceptability, e.g., the particle size of sugar and cocoa in chocolate reflecting the smoothness to the palate, the colour of sauce, the size (and evenness of size) of canned and bottled fruit, the setting power of jelly, the strength of flavour in custard and blancmange and the balance of spices in meat products. On a more long-term basis the texture of sections of bakery products can be photographed and comparisons made at any future date.

During the development of the recipe, it is decided whether permitted additives and processing aids should be incorporated into the food. The factory laboratory must ensure that the correct amount has been added, or keeping qualities, texture, colour, sweetness, etc., will be impaired. Thus with preservatives, for instance, it is necessary

14

to check that the amount present in the final product complies with factory minimum as well as legal maximum limits.

Storage characteristics can be predicted from selected chemical tests. Yeasts and moulds are liable to grow in foods such as candied peel, glacé cherries and preserves unless they contain two-thirds of total sugars in the aqueous phase. Additionally, approximately two- to four-sevenths of the total should be present as invert sugar, otherwise crystallisation is liable to ensue on storage. Also, with lemon curd there should be 3–4% of efficiently gelatinised starch present to prevent fat separation. With foods such as pickles, sauces and dressings, which rely on vinegar rather than sugar or salt for preservation, the product is checked for the sufficiency of volatile acidity in the aqueous phase.

Wrappers protect the product from contamination and dry foods of low equilibrium relative humidity have to be packed in packaging material that is impermeable to moisture vapour. If the same material is used for 'wet' foods, however, condensation and mould growth rapidly arise within the package. 'Open-pack' flesh products are therefore packed in material that is permeable to moisture vapour. To prevent the formation of unsightly dried-up patches on frozen fish, the material is surrounded by either impermeable film or a protective barrier of ice before storing it below 0 °F.

STORAGE CONTROLS

Examples of the prediction of occurrences during the storage of the food have been mentioned in the previous section as certain routine tests on the product are carried out for this purpose. In other instances the 'life' of the product can be assessed from accelerated storage tests, in which the package, can, bottle or jar is submitted to the effect of elevated temperatures (and sometimes humidities also). The container and contents are then examined for faults, rancidity, growth of organisms and insects, loss of colour and so on.

The manufacturer must be concerned also with the way the product is stored after it has left the factory. This is where leaflets of instructions to retailers, wholesalers and customers should assist. The shopkeeper must be reminded of the importance of stock rotation on the 'first-in, first-out' principle—particularly with perishable and frozen foods. With meat products, such information must include advice as to keeping times for the shop that possesses no refrigerator as well as for large supermarkets with large cold-stores. This is a difficult problem, but in view of the importance of constant vigilance at all stages, the message must be repeated several times and discrete checks should be made to ensure that the instructions are being passed 'down the line'.

COMPLAINT SAMPLES

Although the proportion of foods about which complaints are made is extremely small in comparison with the total turnover, each complaint must be treated as being of serious concern. Complaints obviously vary in type, but the commonest are probably 'not the same as usual', 'mouldy', 'gone-off', 'funny taste' or that the food contains insects or some form of extraneous matter. Unsatisfactory taste, odour and appearance often have been caused by over-prolonged or unsuitable storage and, with these types of complaint, code numbers are often helpful. Examples of typical laboratory tests on complaint samples of this type are tests for moisture in cereals and for rancidity of any fats present. The report on such samples should include any code or batch numbers and, if possible, positive advice should be given to try to prevent a recurrence of the trouble. Apportioning blame to individuals seldom helps.

COMPETITORS' SAMPLES

In a competitive market, information about products made by other manufacturers is always of interest to management. Competitors' samples are also valuable when one is proposing to start manufacturing a new product. Organoleptic characteristics are as important as scientific tests and the report should bring out comparisons with one's own product (where applicable) by tabulating the results of all samples on the same report.

REPORTING OF LABORATORY RESULTS

Although a few samples and checks (e.g., in process control) are reported upon verbally, the laboratory is normally expected to submit written reports on its work. In addition, the factory chemist is often expected to inform management by means of brief memoranda of topical scientific matters related to the firm's activities. This particularly applies in the case of new and pending legislation.

The method of reporting varies according to the individual and is usually acquired by a combination of common sense and experience, but what follows is designed to present some basic 'rules' for those who are confronted with the task for the first time. One should always have in mind what is the most useful type of information which can be given to the main person who is likely to act upon the report's findings. The degree of accuracy in expressing any data must also be considered.

Some laboratory work is carried out to give background information or confirmation of results and it is not always necessary to report all the figures obtained. In a somewhat similar manner, only one figure should be returned for each constituent even if replicates have been carried out. A typical howler is the following extract from a student's report:

Lead (sulphide method)   5 ppm $\Big\}$ (max. 2 ppm)
Lead (reference method)   Nil

Firstly, executives are normally interested in the result, not the method used. In fact, with few exceptions, the method is seldom divulged on the report, although it may arise in subsequent correspondence arising from a dispute. Secondly, the sulphide method is a sorting test used to give an approximate idea of the figure. Therefore, in view of its indication of the sample probably exceeding the limit, another procedure was applied to confirm the result. Presuming that one is satisfied about the manner in which the analysis by the longer method was carried out, the result from it must be taken as being the correct one, particularly as it is the 'reference' method. So 'nil' is the only figure to insert on the report. Also, widely differing results such as these should not be 'averaged'!

Most workers agree on the terms used in reports for analytical data, e.g., 'Moisture', 'Ether extract', 'Fibre'. Also, total nitrogen obtained by using the Kjeldahl procedure is reported with the appropriate factor, e.g., as 'Protein ($N \times 6.38$)' in the case of dried milk. Some expressions, however, are used that are more useful to the factory floor. For example, the strength of acid phosphates is reported as the equivalent to bicarbonate (to assist with recipe calculations) rather than as a percentage assay figure. Also, miscalculations are prevented by having a previously prepared table or by merely reporting the titration on the report:

Acidity   3.7 ml 0.1N NaOH/10 g

Such an expression is just as effective from the point of view of interpretation, as control limits can still be based on them.

For the degree of accuracy in reporting it is impossible to lay down rules as exact as those in pure science. For example, accuracy is obviously limited if only one determination is performed. The following, however, are typical of what appears in reports in many different organisations:

(*i*) Percentages above unity are commonly given to the nearest 0·1, e.g., moisture 11·6%, fat 28·7%, protein (N × 6·25) 12·2%.

(*ii*) Percentages below about 1·0 are sometimes given to the second place of decimals, e.g., ash 0·64%, acid-insoluble ash 0·08%.

(*iii*) Preservatives are often reported to the nearest 5 or 10 ppm.

(*iv*) Unless the limit is very low (e.g., soft drinks), trace elements are returned to the nearest whole number. The results of limit tests can be more generally stated as 'less than X ppm'.

In deciding on the degree of accuracy to be used in reporting, it is also most important to take into account interpretation. If we consider, for instance, that the moisture in a cereal received as a raw material should not exceed 13%, there is little advantage (particularly if an electrical meter has been used) in reporting a much lower result such as 10·3% to any nearer than the nearest whole number (10). With trace elements (*iv*, above), if the legal limit is 20 ppm the confidently determined figure of 3·7 ppm can be returned as 4 or less than 5 ppm and still be just as valuable to the person reading the report. The same applies in the case of preservatives from the point of view of legal interpretation, although we may then be working more to lower factory limits based on the recipe than to legal aspects. Also, the meat and fruit contents of products are reported to the nearest whole number, which may, in view of the dubious averages from which they are derived, often be a quite fictitious degree of accuracy.

RECORDING OF SAMPLES RECEIVED AND REPORTING

Each sample received in the laboratory should be given a number and recorded in a master book. It should then be labelled with full particulars, including the date of receipt. It is also useful to allocate a loose-leaf sheet for each sample. The laboratory work can be written on the reverse side and, after this has been completed, the draft of the final report can be written on the front. After typing, the loose-leaf sheet can be put in the files.

Where large numbers of similar examinations are being performed daily on a particular product, much reporting time can be saved by having printed forms that contain printed sections for each measurement and comments. When a large number of different types of samples are being handled daily, it is necessary to have a general basic form to cover all eventualities, as shown opposite.

Sample   (a) .... No.:........,....
        (b) ....
Received from   (c) ....
Sample received on (d) ....
Report submitted on (e) ....
Remarks   (f) ....
....

REPORT

(g) *Examination*

(h)

(i)

(j)

Signature ....

The sections of such a form should be filled in as follows.

(a) Appropriate title, e.g., cornflour, pumping brine, lemon curd. Include brand name if appropriate.

(b) Insert other particulars, e.g., number and size of containers, code numbers, and type of sample ('Production sample', 'Buying sample').

(c) Include any special particulars as to source, e.g., name of wharf, supplier, or individual or department in the factory.

(d)
and
(e) Relevant dates.

(f)   For other points of interest prior to the laboratory examination, e.g., 'Replacement of previous delivery (Report No. 413, 1st July, 1972)', 'Colour questioned by Mr. Jones'. Particulars on the label of a competitor's sample are also appropriate here.

(g)   Insert 'Chemical', 'Microbiological', etc., as appropriate.

(h)   List (and tabulate if more than one sample is being reported upon) the data and other laboratory findings. Report figures clearly as 'per cent', 'ppm', ' μg/oz', 'fluid oz' and put factory control limits or legal standards or limits in brackets after it. Where doubt exists as to what is the desired result of qualitative tests, report as 'Satisfactory' or 'Not satisfactory', rather than as 'Negative' or 'Positive'.

(i)   With satisfactory samples showing no abnormalities or deviations from the norm, this part can be ignored (see j). If the interpretation is not obvious from a study of the figures together with the relevant standards in brackets afterwards, however, some further statement should be made here. Anything different from normal, or where there is doubtful compliance with legislation or specification limits, should be further emphasised in words. This part should always be used if the sample is not satisfactory. Deficiencies should be calculated as percentages of the standard (e.g., the legal minimum).

(j)   With raw materials and most products this section gives a quick summary. The wording is therefore usually 'Sample satisfactory' or 'Sample not satisfactory'.

EXTRACTS FROM LABORATORY REPORTS

I. *Wheat flour, cake grade (raw material)*
   Chemical examination

	per cent
Moisture (Carter–Simon)	18·1
Total ash	0·71
Protein ($N \times 5\cdot7$)	8·7
Amount retained by 120-mesh sieve	0·17

This sample has a higher moisture content than normal (it usually does not exceed 16%). In my opinion, the moisture content of this grade of flour using the Carter–Simon method should preferably not exceed 16%.

II. The figures given in brackets in the next example are those which should be obtained if the correct recipe has been adhered to and the batch mixed adequately.

Baking powder (production sample)
  Chemical examination

	per cent
Total carbon dioxide	11·0 (14·4)
Residual carbon dioxide	0·3 (Nil)
Available carbon dioxide	10·7 (14·4)

Although the amount of available carbon dioxide in this sample is less than usual, it is above the minimum of 8·0% prescribed by The Food Standards (Baking Powder and Golden Raising Powder) Order.

N.B. A batch like this would not be wasted—it might be 'married in' or sent to caterers, who normally use up deliveries in a comparatively short time. Powders that go on to the retail market may not be used for a considerable period, in which time the available CO_2 may well have fallen in this instance below the statutory minimum of 8%.

III. In this example the figure in brackets represents the minimum statutory standard.

Shredded suet (received as raw material)
  Chemical examination

	per cent
Moisture	2·5
Fat	80·3 (83% minimum)
Dry rice flour (by difference)	17·2
Extracted fat	
Iodine value	45·3
Saponification value	201
Refractive index at 60 °C	1·451
Free fatty acids (as oleic acid)	0·1

The Food Standards (Suet) Order requires that shredded suet contains at least 83% of beef fat. This sample contains only 80·3% and is therefore deficient in this ingredient to the extent of 3%.

IV. *Marmalade (production sample)*

Soluble solids (%)	70·6 (69–71)
Reducing sugars (as % invert sugar)	47·0 (20–40)
pH (20% solution)	3·1 (3·0–3·5)
Set	Satisfactory

The proportion of reducing sugars present is considerably above that normally found in our marmalade and it is probable that crystallisation will ensue on storage.

V. *Mint (raw material)*

Total ash	11·1%
Acid-insoluble ash	0·9%
Volatile oil	2·0%
Stalks	7%
Arsenic	7 ppm (max. 5)
Lead	2 ppm (max. 10)

The Arsenic in Food Regulations prescribe that the arsenic content of dried herbs shall not exceed 5 ppm. This sample of mint contains 7 ppm of arsenic, i.e., 2 ppm in excess of the statutory limit.

QUALITY CONTROL CHARTS

Control limits have been applied in industry for a very long time. When recipes were originally worked out they were costed to see that the customer obtains a reasonable product both as regards weight (or volume) and general quality. This requires the application of limits as to weight, composition, etc., at various stages of the manufacturing process. The variations that occur can be attributed to variation in the raw material, the machinery and the operators. Such variations can be visually represented on control charts, which can be used for raw materials, in process control or for the product. This section is not designed to cover all these stages, but the information given should be sufficient for the student to build up his first charts from raw materials received or on products made on the pilot scale. More detailed information can be found by reference to specialist publications[6,15-19]. For a scheme involving quality control charts to be successful it is important to be certain that a reasonable number of measurements of the same type will be available. Industrially, control charts have largely been applied in weight control, but they have obvious applications to container capacity, physical measurements and compositional data. It might be added that as far as weight control is concerned there is more or less universal acceptance of automation in (a) weighing, (b) rejection of lightweights and (c) feed-back of information which results in automatic machine adjustment.

The control chart normally has a central line representing the average quality level (AQL) together with two pairs of lines on each side of it corresponding to the upper and lower action and warning limits for the process. In setting up the chart, the selected attribute is measured on a large number of samples and the frequency is plotted against the values obtained. These should have a normal (Gaussian) distribution and produce the familiar bell-shaped curve

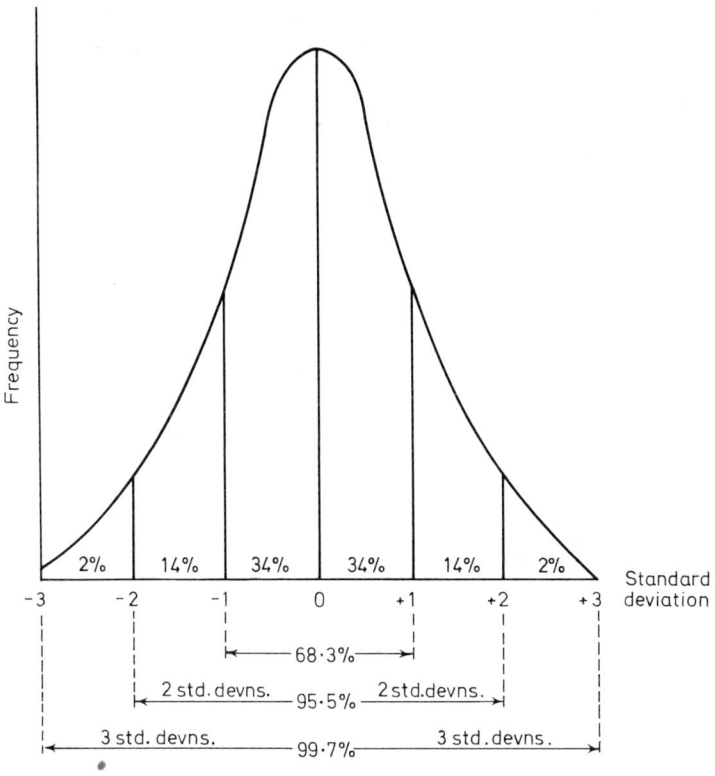

Figure 1.3. Graph illustrating normal distribution

(*Figure 1.3*). The area under the curve between standard deviations $+1$ and -1 covers about two-thirds of the probability that a single sample falls within the range. About 95% fall between $+2$ and -2 and over 99% between $+3$ and -3. Then the *Average Control Chart* for the attribute can be drawn with the warning lines at $+2$ and -2 and the action lines at $+3$ and -3 (*Figure 1.4*). Each point on the average chart represents the mean of the results of several samples taken from the process at the same time. Apart from the mean, however, it is necessary to keep a check on the variability in the process, which is essentially the variation in the standard deviation between each set of samples measured. So we also set up a *Range Control Chart* (also shown in *Figure 1.4*) covering the differences between the minimum and maximum for each set of sample measurements made. Although the placing of the various lines on these

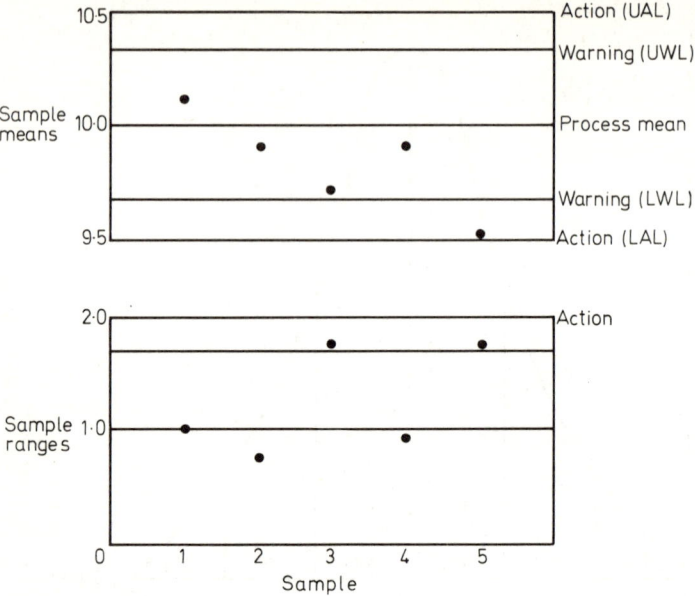

Figure 1.4. Average (upper) and range (lower) control charts used for weights. Each point represents the mean of values obtained by weighing four samples (see *Table 1.2*)

charts should be calculated statistically, the vertical axis is, of course, represented as the actual measurements rather than as standard deviations. Further application to full-scale production often reveals wider variations than were encountered during trials and the action and warning limits may later have to be widened in the light of experience. Also, in the case of the measurement of a composition figure to which a legal quality standard applies, it is important to set the lower action limit above the minimum requirement. If the measurement is based on a relatively lengthy laboratory method, no immediate action can be taken and the data on the chart can only be studied from the point of view of trends over a period. This is not then really true control, which infers that some immediate action can be taken.

For the compilation of *Table 1.2*, the mean values were determined from four samples taken at hourly intervals. The average and range figures were then applied to the charts shown in *Figure 1.4*. When warning limits are approached or exceeded, appropriate adjustments

Table 1.2. WEIGHT MEASUREMENTS USED FOR THE
CONSTRUCTION OF THE CONTROL CHARTS IN FIGURE 5

Sample	1	2	3	4	5	
Weight	10·4	9·6	10·1	9·9	10·1	
	10·4	9·9	10·4	10·3	8·4	
	9·4	9·9	8·4	9·4	10·1	
	10·1	10·3	9·9	9·9	9·4	
						Mean
Average	10·1	9·9	9·7	9·9	9·5	9·82
Range	1·0	0·7	2·0	0·9	1·7	1·20

should be made (samples 3 and 5 on the range chart). If the action limits are exceeded, the process should be stopped for immediate investigation (sample 5 again). It must be appreciated that for convenience total weights (contents + package) are recorded. Due allowance has to be made, however, for the average weight of the containers, which are similarly checked on receipt as a raw material.

REFERENCES

1. BLYTH, A. W. and BLYTH, W. B., *Foods: Their Composition and Analysis*, 5th edn, Griffin, London (1903)
2. BASSETT, C. A., *Chemy. Ind.*, No. 30, 786 (1956)
3. HUGHES, E. B., *Pure Food and Pure Food Legislation; Papers of the 1960 Centenary Conference* (Edited by A. J. Amos), Butterworths, London, 21 (1960)
4. KIDD, F. and WEST, C., *The Refrigerated Gas-storage of Pears*, D.S.I.R. Food Investigation Leaflet No. 12, H.M.S.O., London (1949)
5. MORRIS, T. N., *Principles of Fruit Preservation*, 3rd edn, Chapman & Hall, London (1951)
6. KRAMER, A. and TWIGG, B. A., *Fundamentals of Quality Control for the Food Industry*, 2nd edn, Avi, Westport, Conn., U.S.A. (1966)
7. HEATH, H. B., *Fd Process. Packag.*, **33**, 144 (1964)
8. *Methods of Sampling and Testing of Gelatines*, BS 757: 1959, British Standards Institution, London
9. CAMPBELL, L. E., *J. Soc. chem. Ind., Lond.*, **57**, 413 (1938)
10. KOPROWSKI, W. S., *Analyst, Lond.*, **76**, 732 (1951)
11. GREIG, R. A. and SEAGRAN, H. L., *Comml. Fish Rev.*, **27**, No. 12, 18 (1965)
12. MARTEN, J. F., *Lab. Pract.*, **12**, 134 (1963)
13. COOK, A. H. and HUDSON, J. R., *Lab. Pract.*, **12**, 52 (1963)
14. BEST, S., *Fd Trade Rev.*, **34**, No. 10, 50 (1964)
15. HERSCHDOERFER, S. M. (Editor), *Quality Control in the Food Industry*, Vols. 1–3, Academic Press, London and New York (1967–1972)
16. DUDDING, B. P. and JENNETT, W. J., *Control Chart Technique when Manufacturing to a Specification*, BS 2564:1955, British Standards Institution, London

17. MINISTRY OF DEFENCE, *Sampling Procedures and Tables for Inspection by Attributes*, DEF-131-A, H.M.S.O., London (1965)
18. KALDY, M. S., *Fd Technol., Champaign*, **21**, No. 4, 19 (1967)
19. KALDY, M. S., *Fd Technol., Champaign*, **21**, No. 5, 47 (1967)

ADDITIONAL SUGGESTED READING
Further sections in references 6 and 15 above.
AMERINE, M. A., PANGBORN, R. M. and ROESLER, E. B., *Principles of Sensory Evaluation of Food*, Academic Press, London and New York (1966)
KINGSLAND, R. O., 'Raw Material Quality', *Fd Technol. Aust.*, **16**, 68 (1964)

2

GENERAL METHODS—BASIC CONSTITUENTS

Sampling—Moisture—Equilibrium Relative Humidity—Fat and Volatile Oil—Fibre—Protein—Ash—Sugars—Acidity—pH Value—Alcohol

2.1 SAMPLING

Prior to each analysis, a representative sample of the material must be carefully prepared. Apart from the following general notes, some foods require special preparation and in some instances these are given in later chapters devoted to the particular foodstuff.

Dry foods should normally be passed through an adjustable hand or mechanical grinder and then mixed in a mortar. In some instances it is advisable to pass the powder through a sieve of suitable mesh size.

Quartering, in which two quarters are rejected and the other two are mixed and the process repeated until a sample of suitable size is obtained, is shown diagrammatically in *Figure 2.1*.

Hard foods such as chocolate should be grated.

Moist foods such as meat and fish products and vegetables should be passed through a hand or mechanical mincer and then mixed in a mortar. The process is then repeated at least once more before transferring the mixture to a stoppered jar and storing it under refrigeration.

Wet foods, particularly those that contain fruit and vegetables such as pickles, sauces and canned products, are often best treated in a high-speed blender. Care must be taken, however, with emulsions such as salad cream or cream soups as such blenders are liable to cause fat separation.

Oils that are not clear should be warmed slightly (stearine sometimes separates on cooling). Otherwise, the heated sample should be filtered. Fats should be filtered after melting them. If an antioxidant is present it may be lost if the sample is filtered at too high a temperature.

Fatty emulsions such as butter or margarine should be warmed to 35 °C in a screw-capped jar and shaken.

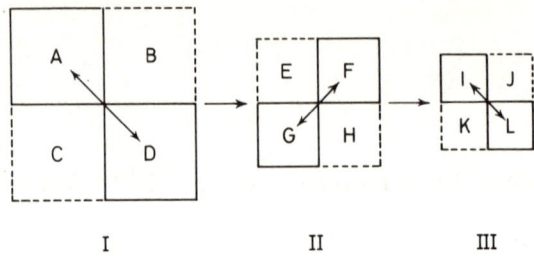

I   II   III

Figure 2.1. Preparation of samples by quartering

I. Powdered sample is spread out in the form of a square and divided into squares. Quarters B and C are rejected. Quarters A and D are mixed to give II
II. Sample is made into the form of smaller square, opposite quarters E and H are rejected and F and G are mixed to give III
III. Process is repeated, J and K are rejected and I and L are mixed, etc., until a sample of suitable size is obtained

In general, prepared samples should be transferred to screw-capped jars and factors such as retardation of spoilage by storage at suitable temperatures and proneness to moisture changes should also be considered.

SAMPLE NUMBERS

A sufficient number of units of the material are drawn from the batch to make up the sample size. The units of material or product drawn are known individually as 'sample units' and collectively as 'a sample'. Often, a composite sample is prepared from a selected number (n) of 'unit' containers. The general expression is

$$n = C\sqrt{N}$$

where N is the population. C is a factor related to the desired degree of accuracy and to the homogeneity. C is less than unity for a homogeneous population, but becomes greater than unity as the heterogeneity increases. With raw materials in particular, the sampling frequency is varied according to such factors as history of material, history of supplier, price and current supply position. For further information on such aspects of sampling the reader should refer to references 6, 15 and 17 on pp. 25 and 26.

2.2. MOISTURE AND TOTAL SOLIDS

In most laboratories in the food industry, moisture is determined in numerous samples every day. Maximum levels are often laid down in commercial specifications. There are several reasons for this, including the following.

(a) The purchaser of raw materials does not wish to buy excessive water.

(b) Water, if present above certain levels, encourages the growth of micro-organisms.

(c) Legal maxima are prescribed for butter, margarine, dried milk and cheese.

(d) The presence of water in powders causes caking, e.g., of sugar and salt.

(e) The level of moisture in wheat has to be adjusted to make it suitable for milling.

(f) The amount of water present may affect the texture, e.g., in cured meats.

(g) The assessment of water content represents a simple way of controlling concentration at various stages of food processing.

The accurate determination of the total water content is often difficult. In practice however, a method that gives good replication so that results are comparable, provided the same procedure is strictly adhered to on each occasion, is usually quite adequate. Some very rapid methods are also available, but the results may have to be calibrated against a more conventional procedure. Apart from the use of refractometers and hydrometers to give indirect measurements of dissolved solids, which are dealt with in other parts of this volume, the main methods for estimating moisture and total solids can be considered under one of the following headings:

(i) drying methods, in which free water is removed by heat or a desiccating agent;

(ii) direct distillation methods;

(iii) rapid electrical methods;

(iv) chemical methods.

(i) DRYING METHODS

Drying methods in which the percentage weight loss of water is calculated, usually after removal by heating under standardised conditions, is the commonest method of estimating the moisture content in foods. Although such methods give consistent results that can be interpreted on a comparative basis, it should be borne

29

in mind that (*a*) it is sometimes difficult to remove all the moisture present by drying, (*b*) at a certain temperature each food is liable to decompose, so that substances other than water are also volatilised, and (*c*) volatile matter other than water may also be lost.

With cereals, the loss in weight due to volatilisation increases as the drying temperature is increased (*Figures 2.2* and *2.3*). From

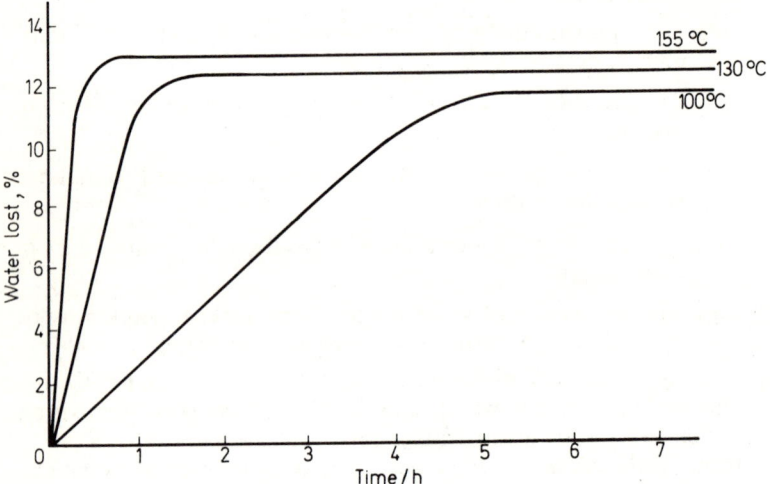

Figure 2.2. Rate of loss of moisture from wheat flour at various drying temperatures

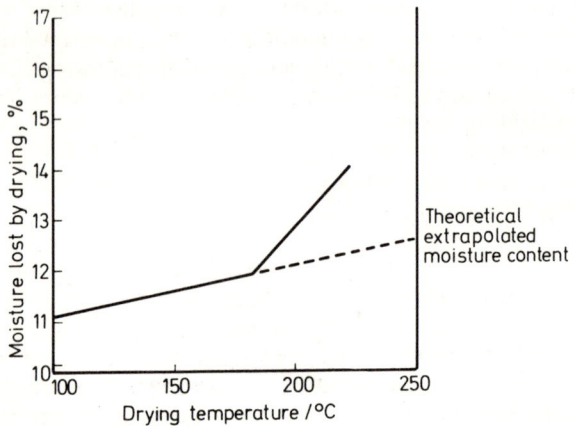

Figure 2.3. Graph showing increase in moisture lost on drying with increase in drying temperature

Figure 2.2 it is apparent that a maximum value is eventually reached at each temperature. The different maximum values attained are most easily explained by considering that at a particular drying temperature cereals contain 'free' and 'bound' water. Although the proportion of water bound decreases as the drying temperature increases, it is extremely difficult to remove all the moisture in the sample. It is more likely, as shown in *Figure 2.3*, that decomposition takes place with flour at about 180 °C as shown by the discontinuity in the straight line. Extrapolation of the main straight line to 250 °C to give the 'true' water content as suggested by Nelson and Hulett[1] is now considered to give low values. The rate of loss of moisture can be increased by drying under reduced pressure. Vacuum drying is particularly helpful with foods that decompose at comparatively low temperatures. For example, foods that contain much sugar and fruit products are preferably dried in a vacuum oven at, say, 70 °C.

If the food is relatively rich in other volatile matter, as with spices, the proportion of interfering constituents is fairly constant and the results are sufficiently accurate for comparative purposes. An alternative method is to dry the material in a vacuum desiccator.

METHOD 2.2A. DETERMINATION OF MOISTURE BY DRYING IN AN OPEN DISH

Place a flat-bottomed, numbered metal dish of about 7 cm diameter (*a*) in a thermostatically controlled oven (*b*) at the selected drying temperature for about 20 min, then cool it in a desiccator and weigh it. Weigh out the recommended amount of sample (*c*) and distribute it over the bottom of the dish. If much water is present (*d*), first partially dry the material on a water-bath before placing it in the oven. Otherwise, transfer the dish containing the sample directly into the oven for the recommended drying time and then place it in a desiccator. Weigh the dish *immediately* it is cold (*e*). For weighing to constant weight, replace the dish in the oven for half-hourly or hourly intervals.

Notes

(*a*) In some instances, sand should be added to the dish to give a greater surface area of drying. A flat-ended rod (*Figure 2.4*) may also be dried with the dish to assist in spreading out the sample.

(*b*) Ovens should preferably be fitted with an internal fan to give an even distribution of temperature and more rapid drying. The fan should be switched off whenever the door is opened.

31

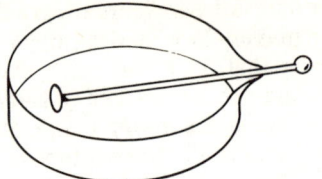

Figure 2.4. Moisture dish with lip and flat-ended stirring rod

(c) With certain exceptions (e.g., gelatine), solid samples must be finely powdered.

(d) With foods such as gelatine, jelly and jam, water is added after weighing out the sample so that the sample dissolves whilst on the water-bath and a more even drying film is obtained.

(e) Prolonged storage of the dried sample in an open dish, even though it is in a desiccator, is liable to cause an increase in weight due to the absorption of moisture. Dried samples should not therefore be left overnight in desiccators before weighing them.

Calculation

$$\text{Moisture (or water) } (\%) = \frac{\text{Weight lost (g)}}{\text{Weight of sample taken (g)}} \times 100$$

$$\text{Total solids } (\%) = 100 - \text{Water } (\%)$$

METHOD 2.2B. DETERMINATION OF MOISTURE BY DRYING IN A DISH FITTED WITH A LID

Place the flat-bottomed, open metal dish in its inverted lid in a thermostatically controlled oven for about 20 min. Transfer the dish and lid to a desiccator, cool them and weigh them together. Weigh out the recommended amount of sample and return the open dish on the inverted lid to the oven. After drying it for the recommended time, replace the lid on the dish, transfer it to a desiccator and weigh the dish immediately it is cold. Most of the notes given under *Method 2.2a* also apply here. Provided the lid fits tightly there is less likelihood, compared with an open dish, of the sample absorbing moisture after it has been heated.

METHOD 2.2C. RAPID DETERMINATION OF MOISTURE USING THE T & M VACUUM MOISTURE TESTER

By using infra-red heating and applying a vacuum, it is possible to remove free moisture from single powdered samples in less than 15 min. In the T & M Vacuum Moisture Tester, the sample is dried on the pan of a torsion balance. The moisture content is read off directly from a drum, which is rotated to bring the balance beam back to the null position after the drying has been completed.

Raise the cover and release the balance pan by pushing back the central clamping lever. Place a 5-g weight on the balance pan, set the drum to 0% and move the zero lever so the shielded balance pointer coincides with the fixed index. Remove the weight and weigh the sample (preferably finely powdered and distributed evenly) on to the balance pan until the balance pointer again coincides with the fixed pointer. Pull down the cover and clamp. Connect the instrument to a vacuum supply, turn the vacuum knob to 'ON' and switch to the appropriate heat (*Table 2.1*). Setting 3 ($\equiv 100$ °C) is suitable for most foods. During the heating, check that the sample is not scorching by looking through the observation window. Rotate the drum so that the balance pointer still coincides with the fixed pointer and read off the percentage moisture lost either after equilibrium has been reached or after a pre-determined time of heating (e.g., 15 min) directly from the scale.

Students should take readings at intervals of 1 min and then draw a graph relating percentage moisture loss *v*. time.

Table 2.1. APPROXIMATE TEMPERATURES OF A 5-g SAMPLE IN THE BALANCE PAN OF THE T & M MOISTURE TESTER AT VARIOUS SETTINGS

Heat control number	Temperature/ °C	Heat control number	Temperature/ °C
1	50	7	190
2	75	8	220
3	100	9	250
4	120	10	280
5	140	11	290
6	160		

METHOD 2.2D. RAPID DETERMINATION OF MOISTURE USING THE CARTER–SIMON RAPID MOISTURE TESTER

The Carter–Simon method for determining moisture is very suitable for industrial control when a series of samples such as cereals have to be examined together. The oven has a drying tunnel, 8 in long, 3 in wide and 2 in high, with hinged flaps at each end to permit the special dishes to be passed through the oven. The oven holds three dishes (each of $2\frac{1}{2}$ in diameter) and during the drying each dish spends an equal period in each of the three positions. The conditions applying to cereals (drying for 15 min at 155 °C) is given in the following method. Different drying conditions may apply, however, with other foods (*Table 2.2*).

Table 2.2. RECOMMENDED CONDITIONS FOR THE DRYING OF VARIOUS FOODSTUFFS IN THE CARTER–SIMON OVEN

Food	Weight of sample/ g	Temperature/ °C	Drying time/ min*
Flours, cereals, starches	5	155	15
Sugar	5	130	9
Butter, margarine	3	130	12
Dried milk	10	130	15
Salt	5	155	9
Meat and fish products	5	135	30

*One-third of this time is spent in each position of the tunnel.

Adjust the oven temperature to 155 °C (cf., *Table 2.2*) after loading the tunnel with three of the special numbered Carter–Simon dishes without lids and each containing about 5 g of wheat flour. Any adjustment of temperature which is necessary is made by turning the screw (anti-clockwise to increase the temperature) under the contact-breaker cover on the top. Whilst the Carter–Simon oven is warming up, dry the dishes and lids in a conventional drying oven (at, say, 100 °C) as in *Method 2.2b*. Then cool the dishes in a desiccator. In each instance, weigh the dried dish in its dried inverted lid and then weigh into it 5 ± 0.005 g of wheat flour (cf., *Table 2.2*). Place the lids in a desiccator in the same order as the dishes will emerge from the oven. Push the first weighed dish (without lid) plus its contents past the flap, reject the emerging 'blank' dish and set the time-clock or stop-watch operating. Exactly 5 min after the first weighed dish was inserted, introduce the next weighed sample through the flap. Continue inserting the 5-g samples every 5 min (again rejecting any 'blank' dishes that emerge) so that each dish plus its contents is dried for exactly 15 min. Place the appropriate lid on each emerging dish containing weighed sample, transfer it to a desiccator, cool it, then weigh it and calculate the loss in weight as a percentage of the original weight of sample.

The lowest temperature recommended for use with foodstuffs is 120 °C, which is normally too high for materials that contain much sugar (cf., *Table 2.2*).

(*ii*) DIRECT DISTILLATION METHODS

Direct distillation methods for determining moisture involve the reflux distillation of the food with a water-immiscible liquid that is lighter than water and usually has a higher boiling point[2,3], e.g., toluene (b.p. 110 °C) or xylene (b.p. 140 °C). The apparatus (*Figure 2.5*) is designed so that both the water and the immiscible liquid volatilise from the flask C, condense in A and fall into the receiver B.

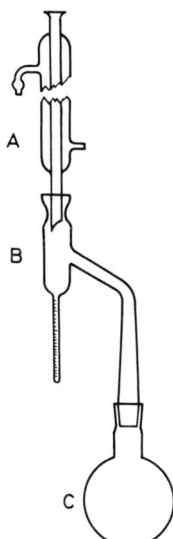

Figure 2.5. Apparatus for determining moisture in foods by direct distillation

The organic liquid mixes with the same liquid in B, and the excess 'refluxes' back into C. The heavier water, however, falls into the graduated tube in the lower part of B and is measured by volume. Solids must be finely powdered.

METHOD 2.2E. DETERMINATION OF WATER BY DIRECT DISTILLATION

Firstly, clean the apparatus thoroughly with a mixture of potassium dichromate and concentrated sulphuric acid and then rinse and dry it. Weigh the sample (e.g., 10 g of flour) into the dry flask, half fill it with toluene (or benzene, xylene or heptane) and attach the condenser and receiver. Boil the liquid by heating the flask with an electric mantle or in an oil-bath until the separated water in the receiver no longer increases in volume (sometimes boiling for several hours is necessary). Coax any droplets remaining in the condenser down into the graduated tube with the assistance of a long, fairly stiff wire (cf., BS 684). Measure the total volume of water (V ml) from the graduations. If W is the weight of sample taken (g):

$$\text{Water in sample } (\%) = \frac{100\,V}{W}$$

Although in practice low results are not uncommon with the distillation method, it has been recommended for the determination of water in ground cereals, oils, soaps and emulsions. It is particularly useful if much volatile oil is present, as with spices. In such instances, the volatile oil either volatilises or mixes with the immiscible liquid in B.

Low results are sometimes due to drops of water clinging to the condenser, even if a long wire is used. If xylene is used, this difficulty can sometimes be overcome by introducing a small volume of ethanol after all the water has been removed from the sample. The distillation method is not suitable for the determination of small amounts of water.

(iii) RAPID ELECTRICAL METHODS

The proportion of water present in foods affects certain electrical properties and this has led to rapid methods based usually on electrical resistance or capacity[4]. In general, instruments based on resistance are not applicable for the determination of very low amounts of moisture, but the relatively simple circuit used makes it the most suitable for portable meters. Those which are based on the dielectric constant can measure moisture down to much lower levels, however. The dielectric constant of water (relative value 80) is much greater than that of most dry materials, so with a moist food its dielectric constant increases as its humidity rises. Circuits that involve capacitance measurements for the detection of the variation in dielectric constant are relatively complex and are preferably used at high frequencies.

The scale readings on meters have to be calibrated against results obtained by other methods (usually oven drying procedures) for each material examined. Such calibrations have been worked out for numerous food materials in the case of the Marconi Moisture Meter (Type TF 933A, based on resistance), the procedure for which is given below. It is suitable for the measurement of moisture contents from 5 to 25%.

METHOD 2.2F. RAPID DETERMINATION OF MOISTURE WITH SPECIAL
REFERENCE TO FLOUR USING THE PORTABLE MARCONI MOISTURE METER

Press upwards on the top boss of the rubber frame, insert the transparent scale for flour, move it along until the air temperature, measured near the cell, coincides with the reference line, then clip it
home. Take the complete cell out of the clamp, remove the metal
plunger, return the upper black ring, and half fill the cavity with the
flour sample. Replace the plunger in the cell and return the cell to the
clamp. Screw up the clamp until the top of the tube mounted on the
end of the screw is flush with the end of the tube. Insert the red and
black sockets in the instrument. Switch to 'Zero' and turn the 'Set
zero' knob until the needle on the meter is central. When setting the
zero, make sure that the 'tens' dial (LH bar-knob) is not set to its
'0' position. Switch to 'Read' and bring the needle to the central
position again by turning the bar knobs. By reference to the 'flour
card', read off the moisture content (in red) against the instrument
reading (in black). The values on the cards and calibration charts
have been standardised against oven drying at 120 °C for 4 h, which
gives results that are close to those obtained using the Carter–Simon
oven.

(iv) CHEMICAL METHODS

There are two chemical methods for determining moisture: (a) the
carbide method[5] and (b) the Karl Fischer titration method[6,7]. In
(a), the powdered sample is shaken with calcium carbide and the
moisture is assessed from the amount of acetylene produced (from
either the volume of gas or the increase in pressure in a closed vessel).
The Karl Fischer method (*Figure 2.6*) is particularly suitable for the
determination of small amounts of water down to about 0.1%.
  The Karl Fischer method applies a reaction due to Bunsen which
can be represented by:

$$SO_2 + I_2 + 2H_2O \rightleftharpoons 2HI + H_2SO_4$$

Fischer used methanol to dissolve the iodine and sulphur dioxide
and added pyridine to the reagent. Although several other liquids
have been suggested[6], methanol is most commonly used for the
extraction of the water from the sample.
  The reaction can be represented as:

$$SO_2 + I_2 + H_2O + 3C_5H_5N \rightarrow 2C_5H_5N\cdot HI + C_5H_5N\overset{SO_2}{\underset{O}{<}}$$

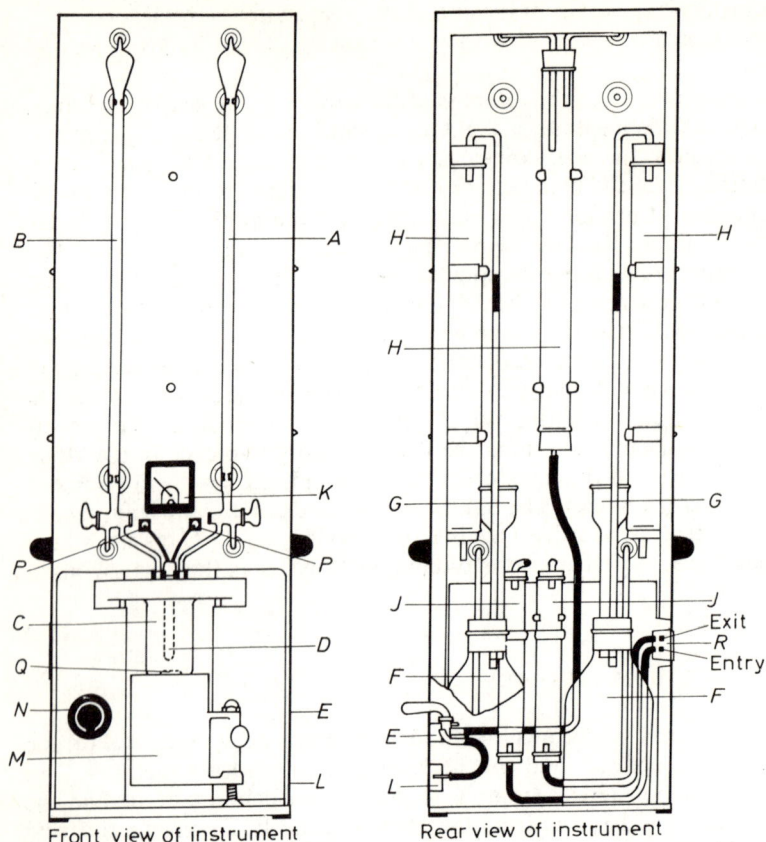

Figure 2.6. BTL Karl Fischer apparatus for the electrometric determination of moisture: A, burette, 25 ml, right side; B, burette, 25 ml, left side; C, titration vessel; D, twin platinum electrode; E, vacuum control stopcock; F, reservoirs for reagents; G, reservoir filling funnels; H, large desiccant guard tubes; J, small desiccant guard tubes; K, end-point indicator; L, vacuum nozzle; M, magnetic stirrer; N, stirrer speed control; P, electrode sockets; Q, stirrer paddles; R, nitrogen drying nozzles. (From Baird & Tatlock Instruction Sheet, No. N. 136. By courtesy of Baird & Tatlock (London) Ltd.)

$$C_5H_5N\diagdown\genfrac{}{}{0pt}{}{SO_2}{\underset{O}{|}} + CH_3OH \rightarrow C_5H_5N\diagdown\genfrac{}{}{0pt}{}{SO_4CH_3}{H}$$

In practice, the equation is not strictly stoichiometric and the Fischer reagent must be standardised against a known weight of water.

For the determination, the water from the sample is extracted with moisture-free methanol and titrated with the Fischer reagent. The end-point can be assessed visually by the brown colour produced due to excess iodine or electrometrically. Commercial instruments using the electrometric procedure are available and in the method described below the Baird & Tatlock Ltd. (BTL) apparatus is used. In view of the sensitivity of the method and reagents, special attention must be paid to detail.

(a) Atmospheric moisture must be excluded from the Fischer reagent and the extracting solvent.

(b) The drying tubes in the apparatus must be packed with an efficient drying agent.

(c) If the water has to be extracted from the sample by heating under a reflux condenser, the liquid should not be subjected to more than gentle simmering[8].

METHOD 2.2G. DETERMINATION OF MOISTURE USING THE BTL KARL FISCHER ELECTROMETRIC APPARATUS (FIGURE 2.6)

Reagents

Karl Fischer reagent Dissolve 63 g of iodine in 110 ml of anhydrous pyridine and immerse the solution in a freezing mixture. Pass in sulphur dioxide slowly, stirring continuously until the weight increases by 32 g. Allow the mixture to stand for 30 min, then dilute it to 500 ml with anhydrous methanol. [1 ml of solution ≡ 5 mg of water (approx.).] The reagent should be standardised daily. The addition of pyridinium hydriodide helps to stabilise the reagent.
Water–methanol standard solution Prepare a solution containing 5 mg of water per millilitre of anhydrous methanol.

Charging the burettes

Remove the back (turn the screw-back until it becomes loose, then pull it off). Check the colour of the silica gel in the tubes. Charge the reservoirs with water–methanol standard solution and Karl Fischer reagent (looking at the *back* of the apparatus, left- and right-hand sides, respectively). Turn the apparatus round so that the KF burette is on the left. Connect the vacuum nipple on the right-hand side to a pump, turn the cock to 'Fill' (pointing outwards), apply suction and fill the burettes via the burette taps. Allow the levels in the burettes to rise above the zero marks, turn the cock to 'Deliver' and set the burettes to zero (ensure that the parts of the burettes from tap to jet are completely filled, as with conventional burettes).

N.B. When the pump is used, the burette cock must not be turned to the 'Deliver' position while the vacuum control cock is in the 'Fill' position.

Standardisation of Karl Fischer reagent

Dry the titration vessel, stirrer and electrode stem by using acetone and a blast of air. Fill both burettes and set them to zero, and switch on the mains electricity. Fit the vessel containing the stirrer on the apparatus and pass a *slow* stream of nitrogen through the entry tube on the left. Add to the titration vessel 20·0 ml of water–methanol standard solution from its burette. The meter should read 'Excess water' (assuming the electrodes are touching the liquid). Switch on the magnetic stirrer and add the Fischer reagent in 1-ml portions at 2-s intervals. Continue to add the reagent until a permanent iodine colour is obtained (titre k ml). The needle should now have swung over to 'Excess Fischer'. Then back-titrate the excess of Fischer reagent with the water–methanol standard solution at 1 drop per second until the needle returns to 'Excess Water' (or gives a steady deflection close to 'Excess water'—a rheostat is provided to adjust the deflection according to the mains voltage. The adjustment should be made with the electrodes immersed in the reaction products after over-titrating with water solution.) Note the *total* volume of standard water–methanol solution added (w ml). Turn off the nitrogen supply, and clean and re-dry the vessel and stirrer.

Assuming that the water–methanol standard solution contains exactly 5 mg of water per millilitre, then 1 ml of Karl Fischer reagent $\equiv \dfrac{5w}{k}$ mg of water.

Alternatively, weigh about 0·1 g of water by difference into the titration vessel and titrate as described above. In this procedure, the electrodes are not immersed in the liquid during the early stages of the titration.

Determination of moisture in food samples

Weigh out rapidly a suitable amount (equivalent to about 20 ml of Fischer reagent) of ground sample into the dried titration vessel, taking special precautions against drying up or moisture pick-up. Add 10 ml of anhydrous methanol and fit the titration vessel with a long, dry air condenser. Heat the vessel by partially inserting it in the top of a heating mantle or on a hot-plate (modified by insulating half of the heating surface with a hardboard sheet so that with a hot-plate temperature of 130 °C the insulated surface is at 70 °C). Bring the liquid to the boil and *gently simmer* it for 15 min[8]. Fit the vessel containing the extract on the apparatus, pass nitrogen through it, titrate it with Fischer reagent and back-titrate it with water–methanol standard solution (as for the standardisation). Carry out a blank determination on 10 ml of methanol.

Students carrying out the method for the first time should compare the results with those obtained by drying methods. Suitable samples are wheat flour, cornflour, boiled sweets, butter, margarine and condensed milk. For the drying of confectionery, weigh the ground sample into a dish containing sand, add water and dry it on the water-bath before drying it in the oven.

The reporting of the results obtained from moisture determinations

Results obtained from determinations of moisture are reported as either 'moisture', 'water' or 'total solids'. There are no really hard and fast rules in any particular instance, but the following comments should represent a guide for the student:

(a) 'Moisture' is mostly used for powders where the amount present is comparatively small, as with flours, cocoa and sugar.

(b) 'Water' is more common when the amount present is rather higher, as with fresh foods, sausages and cheese.

(c) 'Total solids' is more often used for liquids, e.g., vinegar, soft and alcoholic drinks and milk.

In reporting, the writer would normally expect students to indicate the method used, e.g., 'Moisture (Carter–Simon) 13·6%', or 'Moisture (loss on drying at 100 °C) 11·2%'. Some chemists prefer to use more non-committal terms, such as 'Total volatile matter at 100 °C' or 'Loss on drying at 100 °C (water, volatile oil, etc.)'. Usually, however, the main consideration in interpreting the results of moisture determinations is the comparison with values obtained previously by using the same method on the same food.

2.3. EQUILIBRIUM RELATIVE HUMIDITY (ERH)

Relative humidity (RH) is the ratio of water vapour pressure in air to the saturation vapour pressure at the same temperature expressed as a percentage:

$$RH(\%) = (p/p_0)_T \times 100$$

where

p = water vapour pressure at temperature T

p_0 = saturation vapour pressure at the same temperature T

Dew point is the temperature at which saturation occurs when air is cooled without changing its water content.

Water activity (a_w) of a material is the ratio of the vapour pressure exerted by the water contained in the material to the vapour pressure of a free water surface at the same temperature:

$$a_w = (p/p_0)_T$$

where

p = water vapour pressure of the material at temperature T

p_0 = saturation vapour pressure at the same temperature T

Equilibrium relative humidity (ERH) is the relative humidity of the surrounding atmosphere at which the material neither gains nor loses water.

All the above terms are of interest in connection with food storage. If the food temperature falls below the dew point of the ambient air, condensation occurs on the food. The ERH of foods tends to approach the RH of the atmosphere during storage. Dehydrated foods therefore have a low ERH and during storage they absorb moisture asymptotically in order to attain equilibrium with the atmosphere. Conversely, flesh foods with a high water content and an ERH of well over 90% tend to dry up during storage in order to attain a similar equilibrium. The environmental requirements of materials can be assessed from sorption isotherms that relate ERH and the moisture content graphically.

The ERH can be assessed from the changes in weight that occur when weighed samples are held in atmospheres of varying humidity. This can readily be achieved by supporting the food above closed vessels containing either varying concentrations of sulphuric acid or saturated solutions of selected salts.

METHOD 2.3. ASSESSMENT OF EQUILIBRIUM RELATIVE HUMIDITY (ERH)

Into a series of small desiccators, place saturated solutions covering the expected range of ERH (*Table 2.3*) in place of the usual desiccant. Weigh out about 3 g of sample (solid samples should be powdered) into each of a series of air-dried dishes supported in their inverted lids. Transfer each dish to the series of desiccators. After 24 h, remove each dish, wipe the bottom with a dry cloth, place the lid on the top of the dish and weigh it immediately. Return each dish to its desiccator and re-weigh daily. From the results in *Table 2.3* and the changes in weight of the sample after being submitted to the various humidities, estimate the ERH of the sample by interpolation.

Table 2.3. RELATIVE HUMIDITIES OVER SATURATED SOLUTIONS OF VARIOUS SALTS AT $20\ ^\circ$C

Compound	Relative humidity, %
Sodium hydroxide	5·5
Potassium acetate	20
Magnesium chloride	33
Potassium nitrite	45
Magnesium nitrate	56
Ammonium nitrate	64
Sodium nitrite	66
Cupric chloride	68
Potassium iodide	69
Sodium chloride	76
Ammonium chloride	79·5
Ammonium sulphate	81
Potassium bromide	84
Sodium sulphate	93
Potassium nitrate	93
Acid calcium phosphate	95
Potassium sulphate	98

N.B. F. E. M. O'Brien [*J. Scient. Instrum.*, **25**, No. 3, 73 (1948)] lists RH figures for a wide range of salts covering temperatures from 2 °C to 50 °C.

2.4. FAT, ETHER EXTRACT AND VOLATILE OIL

(i) FAT AND ETHER EXTRACT

Fat is usually determined either by direct extraction with a solvent, indirect extraction after treatment with alkali or acid, or by measurement of the volume of fat separated in a graduated tube after mixing it with sulphuric acid or neutral or alkaline reagents and centrifuging the mixture. Reference methods involve weighing the fat. The rapid volumetric methods are found to be very suitable in routine control, particularly when large numbers of samples have to be examined.

Direct extraction of the fat from the *dry* material with light petroleum (boiling range 40–60 °C) is preferred. Diethyl ether is more efficient, but may extract non-fatty materials. The method is most conveniently carried out in a continuous extractor of the Soxhlet, Bolton (*Figure 2.7*) or Bailey–Walker (*Figure 2.8*) type. Direct extraction gives the proportion of free fat.

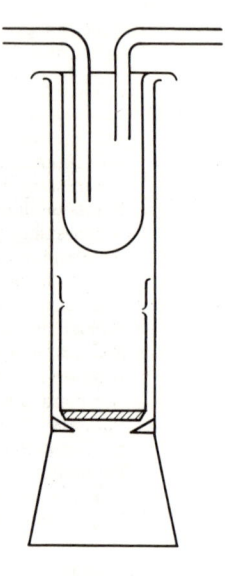

Figure 2.8. Bailey–Walker continuous fat extractor

Figure 2.7. Bolton-type continuous fat extractor

Where protein is liable to interfere with direct extraction, the material is hydrolysed with acid or alkali. Such methods give the free plus combined fat. In the Rose–Gottlieb method (*Method 6.4*, p. 134), alcohol is added to precipitate the protein, which dissolves in ammonia before extraction with mixed ethers. The light petroleum in the solvent reduces the solubility of non-fatty materials such as lactose, which are partially soluble in diethyl ether when it is used alone. In the Werner–Schmid method (*Method 6.20*, p. 159), heating with hydrochloric acid causes the protein to dissolve, prior to extraction with mixed ethers. Treatment of the original material with ammonia prior to the addition of acid encourages the proteins to dissolve more readily. The acid method is less suitable for materials that are rich in sugar. The volumetric Gerber method (*Method 6.3*, p. 132) involves shaking the food with 90% sulphuric acid. A higher concentration may cause charring of the fat and a lower strength is unlikely to release all the fat. Other volumetric processes use less corrosive neutral or alkaline reagents (see Macdonald's method, *Method 6.23*, p. 161) or van Gulik butyrometers (p. 180).

Other techniques involve the measurement of changes of refractive index and specific gravity due to alterations in the concentration of fat in solution.

Details of the main procedures for determining the fat content are described in other appropriate Sections (see Chapters 6 and 7). Most of the basic methods were originally designed for the determination of fat in milk.

(ii) VOLATILE OIL

The determination of volatile oil is most commonly applied to herbs and spices, as it represents a measure of the aroma and to some extent the flavour. When only small samples, such as buying samples, are tested, or when specialist distillation equipment is not available, the volatile oil can be estimated by the direct extraction method. With samples containing relatively little volatile oil, such as fruit juice, squashes and lemon curd, it is advisable to apply the distillation method to a large sample.

METHOD 2.4A. DETERMINATION OF VOLATILE ETHER EXTRACT, FIXED ETHER EXTRACT AND TOTAL ETHER EXTRACT

Weigh out 3 g of spice on a filter-paper, 'wrap round', lightly plug it with defatted cotton-wool and push it down into a continuous extractor of the Soxhlet, Bolton (*Figure 2.7*), or Bailey–Walker (*Figure 2.8*) type. Use diethyl ether for the extraction (*a*). After about 1 h, remove the sample from the extractor, grind it thoroughly in a pestle and mortar and return it to the apparatus to complete the extraction (*a*). Remove most *but not all* of the solvent by heating the flask using a tap-off (see *Figure 6.3*, p. 135) or another suitable method (*b*). Then place the flask containing both fixed and volatile oils and a small volume of diethyl ether on the bench and remove the residual solvent by means of air-bellows. Stop blowing in the air when the smell of ether first disappears, i.e., when only the odour of the volatile oil is discernable, and transfer the flask to a vacuum desiccator overnight. In the morning, weigh the flask containing the fixed plus volatile oils (W_T). Heat the flask and its contents in an oven at 100 °C for 45 min to remove the volatile oil then cool it in a desiccator. Weigh the flask containing the fixed oil (W_F). Clean out the flask, dry it at 100 °C, cool and weigh it (W). Obtain the total ether extract from $W_T - W$, the fixed ether extract from $W_F - W$, and the difference ($W_T - W_F$) represents the volatile ether extract. Express all these results as per cent w/w of the sample (c).

Notes

(*a*) It is most important that the material in the flask never becomes dry at any time during the extraction, otherwise some volatile oil will be lost. In other words, some solvent must always be present in the flask throughout the extraction and until the flask is cooled.

(*b*) If the time factor is unimportant, it is safer to allow the solvent to evaporate off at room temperature.

(*c*) The results from the distillation method are expressed as per cent v/w (see *Method 2.4b*).

METHOD 2.4B. DETERMINATION OF VOLATILE OIL BY DISTILLATION

Thoroughly clean the apparatus (*Figure 2.9*) with chromic acid and introduce into A a suitable amount of sample (*a*) and 300 ml of water (*b*). Connect up the cleaned still-head and run water into F until it overflows at B into the flask (*c*). Insert the 'hollow' stopper G (*d*) and boil the liquid in A by means of an electric mantle or oil bath (*e*) for 3–5 h. Ten minutes after stopping the heating, read off the volume of distilled oil in millilitres and report the results as per cent v/w in the sample (cf., *Method 2.4a*).

Notes

(*a*) Take, say, 20 g of a herb or spice, 250 g of lemon curd. With some foods the sample is introduced later [see Note (*c*) below].

(*b*) By using glycerin or salt solution instead of water the volatile oil is released more rapidly.

(*c*) When the volatile oil to be determined is heavier than water (for example, with anise, fennel, clove, cinnamon, cassia, parsley, pimento and turmeric), also add by pipette exactly 1 ml of xylene or turpentine through F. If this is done, boil the water in the flask first without the sample for 30 min and read off the volume of xylene (V) 10 min after stopping the

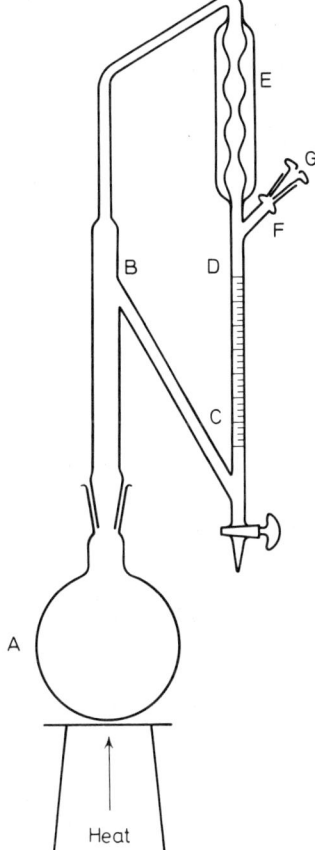

*Figure 2.9. Distillation apparatus
for the determination of volatile oil*

Heat

heating. Then add the sample to A, carry out the distillation
and finally read off the total volume of oil/xylene (V_T) 10 min
after stopping the heating. Then $V_T - V$ gives the volume of
volatile oil.

(*d*) If the stopper has no central hole, place a small strip of
paper in F before inserting G so that the slight gap releases
air if a sudden build-up of pressure arises during the reflux
distillation. The stopper should be *lightly* held in position
with a rubber band.

(*e*) If a burner is used, the flask should be heated on a gauze
with the usual asbestos protecting pad. Direct heat on the
flask may cause localised charring of the sample.

2.5. CRUDE FIBRE

The 'crude fibre' is the washed, dried organic residue that remains after boiling the defatted material successively with dilute sulphuric acid and dilute sodium hydroxide solution. Although the fibre consists largely of cellulose, the amount obtained depends on the analytical procedure used and it is important always to use the same method. In the U.K., a standardised procedure is prescribed in the Fertilisers and Feeding Stuffs Regulations 1968 (S.I. 1968, No. 218), as follows.

METHOD 2.5. DETERMINATION OF FIBRE

Weight (*a*) to the nearest milligram about 2·7–3·0 g of the sample (*b*), transfer the sample to an extraction apparatus (*c*) and extract it with light petroleum (*d*). Alternatively, extract with light petroleum by stirring, settling and decanting three times. Air-dry the extracted sample and transfer it to a dry 1000-ml conical flask (*e*). Add 200 ml of 0·255 N sulphuric acid, this volume being measured at ambient temperature and brought to boiling point, the first 30–40 ml being used to disperse the sample, and heat the mixture to boiling point within 1 min (*f*). An appropriate amount of antifoaming agent (*g*) can be added if necessary. Boil the mixture gently for exactly 30 min (*h*), maintaining a constant volume (*i*) and rotating the flask every few minutes so as to mix the contents and remove particles from the sides. Meanwhile, prepare a Buchner funnel (*j*) fitted with a perforated plate by adjusting a piece of cut cotton cloth or filter-paper to cover the holes in the plate so as to serve as a support for a circular piece of suitable filter-paper. Pour boiling water into the funnel, allow it to remain until the funnel is hot and then drain it off by applying suction. Care should be taken to ensure that the filter-paper used is of such quality that it does not release any paper fibre during this and subsequent washings. At the end of the 30-min boiling period, allow the acidic mixture to stand for 1 min and then pour it immediately into a shallow layer of hot water under gentle suction in the prepared funnel. Adjust the suction so that the filtration of the bulk of the 200 ml is completed within 10 min. Repeat the determination if this time is exceeded. Wash the insoluble matter with boiling water until the washings are free from acid; then wash back into the original flask by means of a wash-bottle containing 200 ml of 0·313 N sodium hydroxide solution, this volume being measured at ordinary temperature and brought to boiling point. Boil the mixture for 30 min with the same precautions as those used in the earlier boiling and treatment. Allow the mixture to stand for 1 min and then filter it immediately through a suitable filter-paper. Transfer the whole of the insoluble material to the filter-paper (k) by means of boiling water, wash it first with boiling water and then with 1% hydrochloric acid (10 ml of concentrated HCl per litre), and finally with boiling water until it is free from acid. Then wash it twice with alcohol and three times with diethyl ether. Transfer the insoluble matter to a dried, weighed, ashless filter-paper (l) and dry it at 100 °C to a constant weight. Incinerate the paper and contents to an ash at a dull-red heat. Subtract the weight of the ash from the increase in weight of the paper due to the insoluble material, and report the difference as fibre.

Notes (not included in the Regulations)

(a) Before weighing out the sample, place the ashless filter-paper required later in its weighing bottle in the drying oven preparatory to cooling in a desiccator and weighing [see also Notes (k) and (l)].

(b) Particle size is important. Solid samples should be passed through a B.S. Test Sieve No. 16.

(c) When the amount of fat or oil is small, the extraction with light petroleum can be omitted[9].

(d) Light petroleum, boiling range 40–60 °C.

(e) Preferably fitted with a reflux water condenser.

(f) With flour, cream 3–5 g of original sample with 20 ml of the cold 0·255 N sulphuric acid before the rest of the 180 ml of (boiling) acid is added.

(g) Frothing can also be reduced by using a 'cold finger' in the flask[10].

(h) Gentle boiling is best achieved by heating the flask with a thermostatically controlled mantle or hot-plate.

(i) Mark the level of the liquid immediately it boils. If a beaker is used instead of a flask it should be covered with a clock-glass and topped up with *boiling* water.

(j) A Hartley[11] divided funnel (see *Figure 10.1*, p. 278) facilitates localisation of the residue in the centre of the paper preparatory to washing off with alkali.

(k) Not the paper prepared at the beginning [Note (a)].

(l) Rather than drying the residue in a filter-paper, the author prefers the method recommended for flour[9], in which the washed, filtered residue is transferred with boiling water to a previously dried and weighed platinum or silica basin.

The water is then evaporated off on a boiling water bath and the dish and residue are dried in the oven at 100 °C. After weighing, the residue can then be ignited in the same dish. Rather similarly, the residue can be filtered on a weighed gooch crucible, which can be dried and ignited.

The outer protective coatings of plant foods such as cocoa and spices contain considerably more fibre than the inner more edible tissues. The fibre is therefore used for assessing the shell content of cocoa and calculating excess husk in white pepper. As the outer tissues of the wheat grain contain more fibre than the starchy endosperm, white flour contains less than does brown flour.

2.6. NITROGEN AND CRUDE PROTEIN

In routine work, the estimation of the total protein is much more frequently carried out than the determination of individual proteins or amino acids. In general (see below), the reference Kjeldahl procedure[12] determines the total nitrogenous matter, which includes non-proteins as well as true proteins. The Kjeldahl method involves the following steps:

$$\text{Total N} \xrightarrow[\text{sulphuric acid}]{\text{Digestion in}} (NH_4)_2SO_4/H_2SO_4$$

$$\downarrow \begin{array}{l} \text{Distil} + \\ \textit{excess} \\ \text{NaOH} \end{array}$$

$$NH_3/\text{Boric acid}$$
$$\text{(titrate with acid)}$$

Sodium sulphate is included in the digestion mixture to raise the boiling point and a catalyst such as copper sulphate is also added to accelerate the reaction. The ammonia in the distillate is trapped either in standard acid and back-titrated, or in boric acid and titrated directly. The Kjeldahl procedure does not, however, determine all forms of nitrogen unless they are suitably modified; this includes nitrate and nitrite.

Most methods of determining nitrogen in foods come under one of the following headings:

(*a*) Macro Kjeldahl distillation.
(*b*) Semi-micro Kjeldahl distillation.
(*c*) Conway micro-diffusion technique.
(*d*) Formol titration.
(*e*) Dye-binding methods.

Methods (*d*) and (*e*) have to be standardised against the reference Kjeldahl method.

The selection of the method for the determination of protein to be used varies according to circumstances, viz., availability of equipment, number of samples to be regularly examined, urgency to obtain results, degree of accuracy called for and the homogeneity of the sample. Thus, if large numbers of milk samples from the same herd of cows have to be examined frequently, the formol titration or one of the dye-binding methods should cope satisfactorily. With a homogeneous powder such as flour, 0·15–0·2 g can be digested prior to rapid semi-micro distillation. The longer reference method is preferable, however, with samples of doubtful homogeneity such as butchers' sausages. On the other hand, in such instances reasonable accuracy should be obtained if, say, 2 g of sample is digested (as in the macro method), the digest is made up to 100 ml and 10-ml portions are taken for semi-micro distillation.

FACTORS USED IN THE KJELDAHL METHOD IN THE ESTIMATION OF CRUDE PROTEIN

The Kjeldahl method estimates the 'crude' protein or total nitrogenous matter. This is calculated by multiplying the total nitrogen (N) by an empirical factor and the result is invariably reported as 'Protein ($N \times 6 \cdot 25$)', 'Protein ($N \times 6 \cdot 38$)', etc. Such factors have been calculated by considering the basic components of a large number of samples of the same food:

$$\text{Factor} = \frac{\text{Mean of total nitrogenous matter by difference}}{\text{Mean of total nitrogen (by Kjeldahl)}}$$

$$= \frac{P_N}{N_K}$$

With meat for instance, taking mean figures,

$$P_N = 100 - (\% \text{ water} + \% \text{ fat} + \% \text{ ash})$$

With other foods, carbohydrates, fibre, etc., must also be subtracted.
The following factors have general recognition:

Meat (also the general factor)	$N \times 6 \cdot 25$
Milk and dairy products	$N \times 6 \cdot 38$
Flour	$N \times 5 \cdot 7$
Gelatine	$N \times 5 \cdot 55$
Egg	$N \times 6 \cdot 68$

METHOD 2.6A. DETERMINATION OF TOTAL NITROGEN BY THE MACRO KJELDAHL METHOD

Special reagents

Catalyst mixture Mix intimately 400 g of sodium sulphate, 16 g of hydrated copper sulphate and 3 g of selenium dioxide.
Screened methyl red indicator Dissolve 0·016 g of methyl red and 0·083 g of bromocresol green in 100 ml of alcohol.

Procedure

N.B. Carry out the procedure in duplicate.

Weigh out (*a*) a suitable quantity (*b*) of the material and transfer it (*c*) to a dry 500–800-ml Kjeldahl digestion flask (*Figure 2.10*). Add 8 g of catalyst mixture and 20–25 ml of concentrated nitrogen-free sulphuric acid (*d*), and mix by swirling. Heat the flask fitted with a

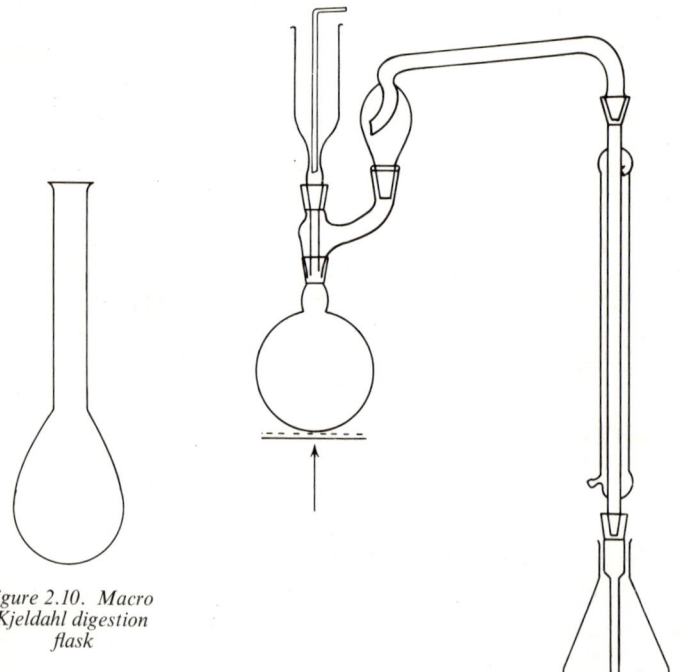

Figure 2.10. Macro
Kjeldahl digestion
flask

Figure 2.11. Macro Kjeldahl distillation apparatus

loose pear-stopper in an inclined position (*e*) in a fume-cupboard. Apply heat gently at first, but when the initial frothing has subsided increase the gas supply gradually until the liquid boils at a moderate rate. Swirl and shake the flask from time to time in order to wash down any charred material adhering to the flask. Continue the heating for 1 h after the liquid has become clear. Allow the flask to cool (*f*). Dilute the mixture with not more than 200 ml of fresh tap water and transfer it to a 1-litre distillation flask (*Figure 2.11*). Wash the mixture in with several small volumes of tap water until the total volume is about 400 ml. To the 500-ml receiving flask add 50 ml of 2% boric acid solution and a few drops of screened methyl red indicator. Add one large piece of granulated zinc to the distillation flask and connect the apparatus to the delivery tube dipping below the boric acid solution. Ensure that all the joints are tight. Add through the tap funnel 75 ml of 50% sodium hydroxide solution, close the funnel and *confirm that the liquid is alkaline after mixing* (*g*). Boil the alkaline liquid in the flask, taking care to prevent undue frothing in the early stages, and distil over about 300 ml (*h*). Open the tap funnel before turning off the gas, wash down the delivery tube into the receiver and titrate the cold distillate with 0·1 N sulphuric acid. A blank should be carried out from time to time. When using reagents of analytical purity grade the blank titration is usually 0·3–0·5 ml of 0·1 N acid.

Notes

(*a*) Unless the material is liquid, weigh out the sample on a folded filter-paper supported on a watch-glass, roll the paper and its contents and drop it directly into the bottom of the flask. A similar filter-paper should be included in the blank.

(*b*) Use an amount of sample containing 0·03–0·04 g of nitrogen.

(*c*) Avoid the material clinging to the neck of the flask.

(*d*) With bulky materials more acid may be required. If this is necessary, more alkali will be required than usual for the distillation.

(*e*) The flask should be supported so that it can be shaken easily. If a special stand is not available, place the flask on a sheet of asbestos with a hole and loosely support the neck in a retort stand ring (not a clamp).

(*f*) The mixture will usually become solid if it is allowed to stand. This can be prevented if the liquid is carefully diluted with a small volume of tap water immediately it cools.

(g) Alkalinity is usually shown by the liquid turning from light to dark blue owing to the effect on the copper catalyst.

(h) Ensure that the rate of flow of the condenser water is sufficient to keep the distillate cold.

Example of calculation

$$\text{Weight of meat product used} = 2.000 \text{ g}$$

Titration

$$= 25{\cdot}5 \text{ ml of } 0{\cdot}1 \text{ N sulphuric acid (blank} = 0{\cdot}5 \text{ ml of } 0{\cdot}1 \text{ N acid)}$$

$$1 \text{ ml of } 0{\cdot}1 \text{ N sulphuric acid} \equiv 0{\cdot}0014 \text{ g of N}$$

$$\text{Total nitrogen} = \frac{(25{\cdot}5 - 0{\cdot}5) \times 0{\cdot}0014 \times 100}{2{\cdot}000} = 1{\cdot}75\%$$

$$\text{Protein} = 1{\cdot}75 \times 6{\cdot}25 = 10{\cdot}9\%$$

METHOD 2.6B. DETERMINATION OF TOTAL NITROGEN BY THE SEMI-MICRO MARKHAM DISTILLATION TECHNIQUE (FIGURE 2.12)

N.B. For reagents required, see *Method 2.6a*. Carry out the procedure on at least two digests from '*a*' or at least two 10-ml aliquots from '*b*', below.

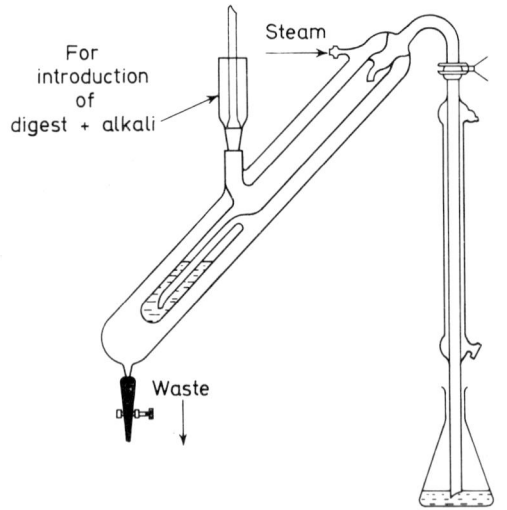

Figure 2.12. Markham semi-micro Kjeldahl distillation apparatus

Digestion

Either (*a*): Weigh accurately about 0·15–0·20 g (more if liquid) of the carefully sampled material into a small digestion flask (*a*). Add 0·8 g of catalyst mixture and 2 ml of concentrated sulphuric acid. Digest the mixture (*b*), heating it gently for about 5 min and then vigorously for 45 min. Cool, dilute the mixture with a few millilitres of water and transfer it to the semi-micro apparatus.

Or (*b*): Digest 2 g as in the macro method, but if possible use less than 20 ml of sulphuric acid. Cool, dilute the mixture with water, transfer it to a 100-ml volumetric flask and make up to the mark at room temperature.

Distillation

Steam out the Markham[13] semi-micro distillation apparatus (*Figure 2.12*) (*c*). Place 10 ml of 2 % boric acid solution and 4 drops of screened methyl red indicator in the 100-ml receiving flask. Transfer quantitatively the diluted contents of the Kjeldahl flask or the 10-ml aliquot to the reaction vessel through the funnel. Rinse out the flask and the funnel with four successive 5-ml volumes of water. Then add 15 ml of 40 % sodium hydroxide solution through the funnel and *confirm that the liquid is alkaline*. If not, add more sodium hydroxide solution. Pass steam through the reaction mixture and distil it for about 15 min with all the pinch-cocks closed. The end of the condenser tube must be below the surface of the boric acid. Finally, lower the receiving flask so that the condenser tube is above the boric acid and continue the distillation for a further 2 min. Rinse the delivery end of the condenser with a small volume of water and titrate with 0·02 N hydrochloric acid. Run off the automatically siphoned-off liquid from the apparatus.

Notes
  (*a*) A 30–50-ml long-necked flask.
  (*b*) A special stand holding six small digestion flasks is available[14].
  (*c*) Other distillation apparatus can be used in a similar manner, e.g., that described by Hoskins[15] or Yuen and Pollard[16].

OTHER METHODS

The Conway[17] micro-diffusion technique can be carried out on a 1-ml aliquot from the macro digestion described in *Method 2.6b*. The formol titration method[18] is given under ice cream (p. 162) and vinegar (p. 257). Dye-binding methods[19] depend on the ability of anionic dyes to combine with protein groups of opposite ionic charge.

2.7. ASH

The ash of a food is an analytical term for the inorganic residue that remains after the organic matter has been burnt away. The ash is not usually the same as the inorganic matter present in the original food as there may be losses due to volatilisation or chemical interaction between the constituents.

The main value of the ash determination (and also the water-soluble ash, the alkalinity of the ash and the acid-insoluble ash) is that it is a simple method for assessing the quality of certain food materials, e.g., a high total ash is undesirable in spices and gelatine. The ash of foods should fall numerically within certain ranges so it also plays a part in identification. Also, sugar and flour can be graded from the ash figure.

The ash is usually determined by ignition as described below, but it is also possible with sugar to assess the ash from the electrical conductivity of its solution. The appropriate factor has to be found separately, however, for each type of sugar examined.

METHOD 2.7A. DETERMINATION OF TOTAL ASH

Hold a clean, flat-bottomed silica dish (approximately 7 cm diameter) in a hot bunsen-burner flame for 1 min, transfer it to a desiccator, then cool and weigh it. Weigh out a suitable quantity (usually 5 g) of the food into the dish and heat it *gently* on a bunsen burner (*a*) in a fume-cupboard until the charred mass is in a suitable condition for transfer (*b*) to a muffle furnace (*c*); continue heating until all the carbon has been burnt away (*d*). Transfer the dish plus ash to a desiccator (*e*), cool and weigh it (*f*). Calculate the total ash as a percentage of the original sample.

Notes
  (*a*) Materials containing much water should first be evaporated on a hot-plate or on a water-bath.
  (*b*) The main consideration is that the material has ceased to give off smoke.
  (*c*) 500 °C is suitable as a general muffle temperature. However, chlorides, etc., may volatilise at this temperature.

(d) With some foods it is difficult to burn away all the carbon. The ashing can be assisted, however, by breaking up the larger particles with a platinum wire. Alternatively, the ash may be treated with water and filtered through an ashless filter-paper. The residue on the paper can then be returned to the dish and burnt off at a higher temperature. The filtrate is then also evaporated in the dish and the ashing completed by a further period in the muffle furnace.

(e) Care is required when the ash is transferred to the desiccator as materials such as gelatine give fluffy ashes that are easily blown away. It is advisable to cover such ashes with a watch-glass or Petri dish immediately after removal from the muffle.

(f) The ash is sometimes utilised for the determination of individual constituents, e.g., chloride, phosphate, calcium and iron.

METHOD 2.7B. DETERMINATION OF WATER-SOLUBLE ASH

Add about 25 ml of water to the ash, cover the mixture with a watch-glass to prevent spurting and boil it gently for 5 min. Filter the mixture through an ashless filter-paper and wash the residue thoroughly with hot water. If the alkalinity of the soluble ash is to be determined, retain all the filtrate. Incinerate the filter-paper in the original dish, cool the ash in a desiccator and weigh it. Calculate the water-insoluble ash as a percentage of the original sample. Then,

$$\text{Water-soluble ash}(\%) = \text{Total ash}(\%) - \text{Water-insoluble ash}(\%)$$

Apart from the figure being a part of the identification, the water-soluble ash, if low, may indicate that the original material has been extracted, as with exhausted ginger and spent tea.

METHOD 2.7C. DETERMINATION OF ALKALINITY OF THE SOLUBLE ASH

Titrate the cooled filtrate from *Method 2.7b* with 0·1 N sulphuric acid using methyl orange as indicator.

The alkalinity is usually calculated on the original sample as either K_2O, K_2CO_3 or Na_2CO_3. The alkalinity of the water-soluble ash is mainly a measure of the alkali carbonates present. It is, in general, an indication of adulteration. Thus, spent tea gives a low figure and alkalisation of cocoa raises the alkalinity of the soluble ash.

METHOD 2.7D. DETERMINATION OF ACID-INSOLUBLE ASH

Moisten the ash with concentrated hydrochloric acid, evaporate it to dryness, bake it to render the silica insoluble and then extract it repeatedly with hot, dilute hydrochloric acid (25 ml of concentrated HCl per 100 ml). Pass the mixture through an ashless filter-paper and wash the insoluble matter thoroughly with hot water. Incinerate the filter-paper in the original dish in the muffle furnace, cool it in a desiccator and weigh it. Calculate the residue as a percentage of the original sample and report the result as 'acid-insoluble ash' or 'sand and other silicious matter'.

The acid-insoluble ash is a measure of the sandy matter present and maxima are specified for herbs and spices. The presence of dirt increases the figure obtained.

METHOD 2.7E. DETERMINATION OF SULPHATED ASH

Ignite the sample *gently* in a silica basin. Cool, moisten with concentrated sulphuric acid and ignite gently again before completing the heating at 800 °C.

The sulphation reduces losses by volatilisation and assists with the attainment of complete combustion. The process converts the fusible and more volatile substances into the more fixed sulphates and therefore produces greater consistency in the ash composition and makes the figure obtained less dependent on the ignition temperature.

2.8. SUGARS

In general, sugars are determined by indirect physical (refractometer, polarimeter or hydrometer) or by semi-empirical chemical (volumetric or gravimetric) methods.

REFRACTOMETRIC METHODS

The indirect assessment of sugars by refractometer is commonly used in rapid factory control. Sugars in solution (often referred to as the *soluble solids*) are expressed as the equivalent of sucrose (*Tables 2.4* and *2.4A*), although Macara[20] has given corrections that can be used for invert sugar, glucose solids and citric acid if greater accuracy is required (see also p. 248).

Table 2.4. REFRACTIVE INDICES* OF SOLUTIONS OF SUCROSE AT 20 °C.

Sucrose, % w/w	Refractive index, n_D^{20}	Sucrose, % w/w	Refractive index, n_D^{20}
0	1·33299	43	1·4056
1	1·33443	44	1·4076
2	1·33588	45	1·4096
3	1·33733	46	1·4117
4	1·33880	47	1·4137
5	1·34027	48	1·4158
6	1·34176	49	1·4179
7	1·34326	50	1·42008
8	1·34477	51	1·42219
9	1·34629	52	1·42432
10	1·34783	53	1·42646
11	1·34937	54	1·42862
12	1·35093	55	1·43080
13	1·35250	56	1·43299
14	1·35408	57	1·43520
15	1·35567	58	1·43742
16	1·35728	59	1·43966
17	1·35890	60	1·44192
18	1·36053	61	1·44420
19	1·36218	62	1·44649
20	1·36384	63	1·44879
21	1·36551	64	1·45112
22	1·36719	65	1·45346
23	1·36888	66	1·45581
24	1·37059	67	1·45819
25	1·3723	68	1·46058
26	1·3740	69	1·46299
27	1·3758	70	1·46541
28	1·3775	71	1·46786
29	1·3793	72	1·47032
30	1·3811	73	1·47279
31	1·3829	74	1·47529
32	1·3847	75	1·47780
33	1·3865	76	1·48033
34	1·3883	77	1·48288
35	1·3902	78	1·48544
36	1·3920	79	1·48803
37	1·3939	80	1·49063
38	1·3958	81	1·49325
39	1·3978	82	1·49589
40	1·3997	83	1·49854
41	1·4016	84	1·50121
42	1·4036	85	1·50391

* International Scale (1936–1959).

Table 2.4A. CORRECTIONS FOR DETERMINING THE PERCENTAGE OF SUCROSE IN SUGAR SOLUTIONS BY REFRACTOMETER WHEN READINGS ARE MADE AT TEMPERATURES OTHER THAN 20 °C. (By courtesy of International Temperature Correction Table, 1936; *Int. Sug. J.*, 1937, **39**, 24s.)

Temperature/ °C	Sucrose, %										
	0	5	10	15	20	25	30	40	50	60	70
	Subtract from the percentage sucrose										
10	0·50	0·54	0·58	0·61	0·64	0·66	0·68	0·72	0·74	0·76	0·79
11	0·46	0·49	0·53	0·55	0·58	0·60	0·62	0·65	0·67	0·69	0·71
12	0·42	0·45	0·48	0·50	0·52	0·54	0·56	0·58	0·60	0·61	0·63
13	0·37	0·40	0·42	0·44	0·46	0·48	0·49	0·51	0·53	0·54	0·55
14	0·33	0·35	0·37	0·39	0·40	0·41	0·42	0·44	0·45	0·46	0·48
15	0·27	0·29	0·31	0·33	0·34	0·34	0·35	0·37	0·38	0·39	0·40
16	0·22	0·24	0·25	0·26	0·27	0·28	0·28	0·30	0·30	0·31	0·32
17	0·17	0·18	0·19	0·20	0·21	0·21	0·21	0·22	0·23	0·23	0·24
18	0·12	0·13	0·13	0·14	0·14	0·14	0·14	0·15	0·15	0·16	0·16
19	0·06	0·06	·0·06	0·07	0·07	0·07	0·07	0·08	0·08	0·08	0·08
	Add to the percentage sucrose										
21	0·06	0·07	0·07	0·07	0·07	0·08	0·08	0·08	0·08	0·08	0·08
22	0·13	0·13	0·14	0·14	0·15	0·15	0·15	0·15	0·16	0·16	0·16
23	0·19	0·20	0·21	0·22	0·22	0·23	0·23	0·23	0·24	0·24	0·24
24	0·26	0·27	0·28	0·29	0·30	0·30	0·31	0·31	0·31	0·32	0·32
25	0·33	0·35	0·36	0·37	·0·38	0·38	0·39	0·40	0·40	0·40	0·40
26	0·40	0·42	0·43	0·44	0·45	0·46	0·47	0·48	0·48	0·48	0·48
27	0·48	0·50	0·52	0·53	0·54	0·55	0·55	0·56	0·56	0·56	0·56
28	0·56	0·57	0·60	0·61	0·62	0·63	0·63	0·64	0·64	0·64	0·64
29	0·64	0·66	0·68	0·69	0·71	0·72	0·72	0·73	0·73	0·73	0·73
30	0·72	0·74	0·77	0·78	0·79	0·80	0·80	0·81	0·81	0·81	0·81

CLARIFICATION OF SOLUTIONS FOR SUGAR ANALYSIS

It is usually desirable to determine sugar in solutions that are clear and transparent. This is especially necessary preparatory to polarimetry. The use of the main clearing agents is discussed below. In some instances corrections for the volume of precipitate formed may have to be made.

Clearing agents

Zinc ferrocyanide (Carrez solution)
    No. 1: 21·9 g of crystallised zinc acetate plus 3 ml of glacial acetic acid per 100 ml of water.
    No. 2: 10·6 g of cryallised potassium ferrocyanide per 100 ml of

water. In use, 5 ml (or more, according to the material and total volume of solution) of No. 1 is accurately added to the solution to be cleared and mixed well. Then an equal volume of No. 2 is added with mixing and, after making up to the mark with water, the solution is shaken well, and filtered after allowing it to stand for at least 20 min. Zinc ferrocyanide is most frequently used for clarifying solutions from milk products.

Alumina cream Prepare a cold, saturated solution of alum in water and make alkaline by stirring-in ammonia solution (sp. gr. 0·88). After allowing the precipitate to settle, wash it by decantation until the wash water gives only a slight precipitate on adding hydrochloric acid and barium chloride. Pour the residual cream into a bottle and stopper it. Alumina cream is sometimes used for precipitating excess of lead as sulphate when basic lead acetate is used as the clarifying agent.

Basic lead acetate Activate litharge (PbO) by heating it for 3 h at 650 °C to produce a yellow solid after cooling. Boil 430 g of neutral lead acetate and 130 g of activated litharge with 1 litre of water for $\frac{1}{2}$ h. Cool and dilute the supernatant liquor to a sp. gr. of 1·25 with recently boiled water. Although effective for clarification, basic lead acetate is liable to introduce errors due to the volume of precipitate, which also may absorb sugars such as laevulose and dextrose. Errors due to the precipitate volume are less apparent if solid basic lead acetate is added instead of the solution.

Neutral lead acetate A 10% or a saturated solution of neutral lead acetate is as efficient as the basic form as a clarifier, but is less prone to absorb sugars.

Sodium tungstate A 12% solution of sodium tungstate has been recommended for use, especially with cereals.

Phosphotungstic acid A 20% solution of phosphotungstic acid has also been recommended for use with cereals.

Animal charcoal In use, the solution is poured through a filter-paper containing decolorising carbon (animal charcoal). This is less efficient than other methods for clarification, but there are no errors due to precipitate volume. Owing to the possible absorption of sugars the first runnings should be rejected.

VOLUMETRIC METHODS

Volumetric methods for determining sugars usually involve the use of alkaline copper solutions, which are reduced to cuprous oxide, or mild oxidising agents such as chloramine T, which react with aldoses. The procedure of Lane and Eynon[21] has largely replaced other copper reduction methods in studies on foods. This procedure involves the determination of the volume of sugar solution required to reduce either 10 ml or 25 ml of mixed Fehling's solution, using methylene blue as the internal indicator. Air is excluded from the reaction mixture by keeping the liquid boiling throughout the titration. Sucrose must, of course, be hydrolysed to invert sugar (dextrose + fructose) before it can be titrated (see preparation of standardised invert sugar solution, p. 64).

METHOD 2.8A. DETERMINATION OF SUGARS BY LANE AND EYNON'S VOLUMETRIC PROCESS[21].

Apparatus (Figure 2.13)

A 50-ml burette, B, is fitted with a pinch-cock instead of a tap and a bent piece of glass tubing, T, so that the burette is not directly over the 250-ml conical flask, F, during the titrations. Glass stopcocks in conventional burettes tend to jam owing to the effects of heat.

Fehling's solution

Soxhlet's modification prepared by thoroughly mixing together equal volumes (accurately by pipette) of A and B—
   A—69·278 g of copper sulphate (CuSO$_4$·5H$_2$O) per litre of water. Filter.
   B—346 g of Rochelle salt (KNaC$_4$H$_4$O$_6$·4H$_2$O) + 100 g of sodium hydroxide per litre of water. Filter.

Methylene blue

A 1% aqueous solution.

Preliminary titration

The solution, inverted if necessary, should normally be cleared first with zinc ferrocyanide (p. 60) and the concentration should be adjusted to give a titre between 15 and 50 ml (0·1–0·3 g of sugar per 100 ml for 10 ml of Fehling's solution; 0·25–0·8 g of sugar per 100 ml for 25 ml of Fehling's solution).
   Fill the burette, B (*Figure 2.13*), with the sugar solution and run a

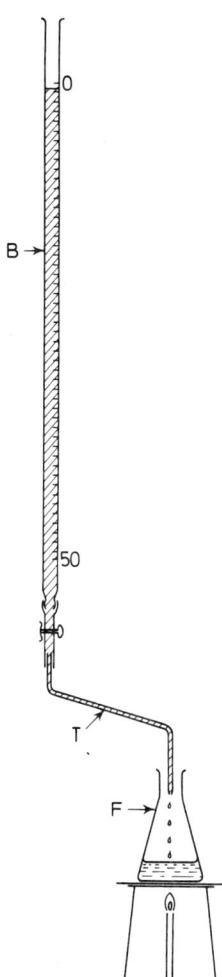

Figure 2.13. Burette for the Lane and Eynon process

small volume out rapidly so as to remove air gaps from T. Pipette 10 ml (or 25 ml) of mixed Fehling's solution into flask F and add 10–15 ml of the sugar solution from the burette. Heat the flask on a plain wire gauze supported on a tripod stand so that the liquid boils briskly (*a*). Then add 4 drops of 1% methylene blue and make further additions of the sugar solution (*b*) at a rate of about 1 ml per 15 s. Keeping the liquid boiling throughout, continue the additions until the blue colour is discharged (titre = *A* ml) (*c*).

63

Accurate titration

Pipette 10 ml (or 25 ml) of mixed Fehling's solution into F and add $(A-1)$ ml of sugar solution from the burette. Swirl the flask round, heat it on the gauze and, after boiling the liquid moderately for $1\frac{1}{2}$–2 min, add 4 drops of methylene blue solution (*a*). Add the sugar solution at a rate of about 0·25 ml per 15 s and complete the titration within 3 min from the time boiling commences (*c*). From the appropriate column in *Table 2.5*, calculate the concentration of sugar in the solution and hence in the sample (see example below). Note that the factors for invert sugar vary according to the proportion of sucrose present in the solution (see *Table 2.5*).

Notes

(*a*) The boiling must be continued *throughout* the titration.

(*b*) If the copper sulphate is completely reduced, as shown by the disappearance of the blue colour, repeat the titration using a more dilute solution of the sample, or if 10 ml of Fehling's solution has been taken, use 25 ml instead.

(*c*) At the end of the titration, pour away the titrated liquid, dissolve the deposit in the flask in dilute hydrochloric acid, and wash it out with water.

Standardisation of Fehling's solution

Dissolve 2·375 g of sucrose (dried at 100 °C) in about 130 ml of water in a 300-ml beaker, add 15 ml of 1 N hydrochloric acid and boil the mixture for 2 min. Cool, add 2 drops of phenolphthalein solution, just neutralise the mixture with 10% sodium hydroxide, transfer it carefully (down a rod) into a 500-ml volumetric flask with water and dilute it to the mark. Mix and titrate the solution by the 'Accurate titration' procedure (above) using 10 ml of Fehling's solution:

1 ml of standard inverted sugar solution $\equiv$ 0·00475 g of sucrose

$\equiv$ 0·005 g invert sugar

10 ml of Fehling's solution

$\equiv$ 10·5 ml of standard invert sugar solution

Calculation of sugar content using Lane and Eynon's method

The proportion of reducing sugars in solution can be obtained from the appropriate factor in *Table 2.5*:

64

Table 2.5. FACTORS FOR LANE AND EYNON PROCESS. (a) Using 10 ml of Fehling's solution

Titre	Invert sugar, no sucrose	Invert sugar +1 g sucrose per 100 ml	Invert sugar +5 g sucrose per 100 ml	Invert sugar +10 g sucrose per 100 ml	Invert sugar +25 g sucrose per 100 ml	Dextrose	Laevulose	Anhydrous maltose	Hydrated maltose	Anhydrous lactose	Hydrated lactose
15	50·5	49·9	47·6	46·1	43·4	49·1	52·2	77·2	81·3	64·9	68·3
16	50·6	50·0	47·6	46·1	43·4	49·2	52·3	77·1	81·2	64·8	68·2
17	50·7	50·1	47·6	46·1	43·4	49·3	52·3	77·0	81·1	64·8	68·2
18	50·8	50·1	47·6	46·1	43·3	49·3	52·4	77·0	81·0	64·7	68·1
19	50·8	50·2	47·6	46·1	43·3	49·4	52·5	76·9	80·9	64·7	68·1
20	50·9	50·2	47·6	46·1	43·2	49·5	52·5	76·8	80·8	64·6	68·0
21	51·0	50·2	47·6	46·1	43·2	49·5	52·6	76·7	80·7	64·6	68·0
22	51·0	50·3	47·6	46·1	43·1	49·6	52·7	76·6	80·6	64·6	68·0
23	51·1	50·3	47·6	46·1	43·0	49·7	52·7	76·5	80·5	64·5	67·9
24	51·2	50·3	47·6	46·1	42·9	49·8	52·8	76·4	80·4	64·5	67·9
25	51·2	50·4	47·6	46·0	42·8	49·8	52·8	76·4	80·4	64·5	67·9
26	51·3	50·4	47·6	46·0	42·8	49·9	52·9	76·3	80·3	64·5	67·9
27	51·4	50·4	47·6	46·0	42·7	49·9	52·9	76·3	80·2	64·4	67·8
28	51·4	50·5	47·7	46·0	42·7	50·0	53·0	76·1	80·1	64·4	67·8
29	51·5	50·5	47·7	46·0	42·6	50·0	53·1	76·0	80·0	64·4	67·8
30	51·5	50·5	47·7	45·9	42·5	50·1	53·2	76·0	80·0	64·4	67·8
31	51·6	50·6	47·7	45·9	42·5	50·2	53·2	75·9	79·9	64·4	67·8
32	51·6	50·6	47·7	45·9	42·4	50·2	53·3	75·9	79·9	64·4	67·8
33	51·7	50·6	47·7	45·8	42·3	50·3	53·3	75·8	79·8	64·4	67·8
34	51·7	50·6	47·7	45·8	42·2	50·3	53·4	75·8	79·8	64·4	67·9
35	51·8	50·7	47·7	45·8	42·2	50·4	53·4	75·7	79·7	64·5	67·9
36	51·8	50·7	47·7	45·7	42·1	50·4	53·5	75·6	79·6	64·5	67·9
37	51·9	50·7	47·7	45·7	42·0	50·5	53·5	75·6	79·6	64·5	67·9
38	51·9	50·7	47·7	45·7	42·0	50·5	53·6	75·5	79·5	64·5	67·9
39	52·0	50·8	47·7	45·6	41·9	50·6	53·6	75·5	79·5	64·5	67·9
40	52·0	50·8	47·7	45·6	41·8	50·6	53·7	75·4	79·4	64·6	67·9
41	52·1	50·8	47·7	45·6	41·8	50·7	53·7	75·4	79·4	64·6	68·0
42	52·1	50·8	47·7	45·5	41·7	50·7	53·8	75·3	79·3	64·6	68·0
43	52·2	50·8	47·7	45·5	41·6	50·8	53·8	75·3	79·3	64·6	68·0
44	52·2	50·9	47·7	45·4	41·5	50·8	53·9	75·2	79·2	64·6	68·0
45	52·3	50·9	47·7	45·4	41·4	50·9	53·9	75·2	79·2	64·7	68·1
46	52·3	50·9	47·7	45·4	41·4	50·9	53·9	75·1	79·1	64·7	68·1
47	52·4	50·9	47·7	45·3	41·3	51·0	53·9	75·1	79·1	64·8	68·2
48	52·4	50·9	47·7	45·3	41·2	51·0	54·0	75·1	79·1	64·8	68·2
49	52·5	51·0	47·7	45·2	41·1	51·0	54·0	75·0	79·0	64·8	68·2
50	52·5	51·0	47·7	45·2	41·0	51·1	54·0	75·0	79·0	64·9	68·3

(b) Using 25 ml of Fehling's solution

Titre	Invert sugar, no sucrose	Invert sugar +1 g sucrose per 100 ml	Dextrose	Laevulose	Anhydrous maltose	Hydrated maltose	Anhydrous lactose	Hydrated lactose
15	123.6	122.6	120.2	127.4	197.8	208.2	163.9	172.5
16	123.6	122.7	120.2	127.4	197.4	207.8	163.5	172.1
17	123.6	122.7	120.2	127.5	197.0	207.4	163.1	171.7
18	123.7	122.7	120.2	127.5	196.7	207.1	162.8	171.4
19	123.7	122.8	120.3	127.6	196.5	206.8	162.5	171.1
20	123.8	122.8	120.3	127.6	196.2	206.5	162.3	170.9
21	123.8	122.8	120.3	127.7	195.8	206.1	162.0	170.6
22	123.9	122.9	120.4	127.7	195.5	205.8	161.8	170.4
23	123.9	122.9	120.4	127.8	195.1	205.4	161.6	170.2
24	124.0	122.9	120.4	127.8	194.8	205.1	161.5	170.0
25	124.0	123.0	120.5	127.9	194.5	204.8	161.4	169.9
26	124.1	123.0	120.6	127.9	194.2	204.4	161.2	169.7
27	124.1	123.0	120.6	128.0	193.9	204.1	161.0	169.5
28	124.2	123.1	120.7	128.0	193.6	203.8	160.8	169.3
29	124.2	123.1	120.7	128.1	193.3	203.5	160.7	169.2
30	124.3	123.1	120.8	128.1	193.0	203.2	160.6	169.0
31	124.3	123.2	120.8	128.1	192.8	202.9	160.5	168.9
32	124.4	123.2	120.8	128.2	192.5	202.6	160.4	168.8
33	124.4	123.2	120.9	128.2	192.2	202.3	160.2	168.6
34	124.5	123.3	120.9	128.3	191.9	202.0	160.1	168.5
35	124.5	123.3	121.0	128.3	191.7	201.8	160.0	168.4
36	124.6	123.3	121.0	128.4	191.4	201.5	159.8	168.2
37	124.6	123.4	121.1	128.4	191.2	201.2	159.7	168.1
38	124.7	123.4	121.2	128.5	191.0	201.0	159.6	168.0
39	124.7	123.4	121.2	128.5	190.8	200.8	159.5	167.9
40	124.8	123.4	121.2	128.6	190.5	200.5	159.4	167.8
41	124.8	123.5	121.3	128.6	190.3	200.3	159.3	167.7
42	124.9	123.5	121.4	128.6	190.1	200.1	159.2	167.6
43	124.9	123.5	121.4	128.7	189.8	199.8	159.2	167.6
44	125.0	123.6	121.5	128.7	189.6	199.6	159.1	167.5
45	125.0	123.6	121.5	128.8	189.4	199.4	159.0	167.4
46	125.1	123.6	121.6	128.8	189.2	199.2	159.0	167.4
47	125.1	123.7	121.6	128.9	189.0	199.0	158.9	167.3
48	125.2	123.7	121.7	128.9	188.9	198.9	158.8	167.2
49	125.2	123.7	121.7	129.0	188.8	198.7	158.8	167.2
50	125.3	123.8	121.8	129.0	188.7	198.6	158.7	167.1

$$\frac{\text{Factor} \times 100}{\text{Titre (ml)}} = \text{Sugar (mg) in 100 ml solution}$$

66

GENERAL METHODS—BASIC CONSTITUENTS

$$\text{Sugar (mg) per 100 ml of solution} = \frac{\text{Factor} \times 100}{\text{Titration (ml)}} \quad (2.1)$$

Sucrose, of course, must be converted to invert sugar before the titration, and

$$\text{Sucrose concentration} = \text{Invert sugar} \times 0.95 \quad (2.2)$$

These relationships are covered in the following example relating to a sample of mixed peel, which contains both sucrose and reducing sugars.

A 10-g sample of mixed peel was macerated with water and made up to 250 ml (4% solution).

Before inversion—
25 ml of 4% solution diluted to 100 ml (1%)
25 ml of Fehling's solution required 24·8 ml of 1% solution
From *Table 2.5*, Factor = 124·0
From equation (2.1),

$$\text{Invert sugar} = \frac{124 \cdot 0 \times 100}{24 \cdot 8}$$

$$= 500 \text{ mg per 100 ml} = 0.50\% \text{ w/v}$$

Invert sugar in sample (%)

$$= \frac{\text{Invert sugar in solution (\%)}}{\text{Sample in solution (\%)}} \times 100$$

$$= \frac{0 \cdot 50}{1 \cdot 0} \times 100 = 50 \cdot 0\% \text{ w/v}$$

After inversion—

25 ml of 4% solution inverted with acid and diluted to 100 ml (1% solution)
25 ml of Fehling's solution required 20·8 ml of 1% solution
From *Table 2.5*, Factor = 123·8
From equation (2.1),

Total sugars (as invert sugar)
$$= \frac{123 \cdot 8 \times 100}{20 \cdot 8} = 595 \text{ mg per 100 ml} = 0.595\% \text{ w/v}$$

Total sugar in solution (%)

$$= \frac{\text{Total sugar in solution (\%)}}{\text{Sample in solution (\%)}} \times 100$$

$$= \frac{0 \cdot 595}{1 \cdot 0} \times 100 = 59 \cdot 5\% \text{ (as invert sugar)}$$

Sucrose (as invert sugar) 67 = 59.5 − 50.0 = 9.5%
From eqn 2.2 sucrose = 9.5 × 0.95 = 9.0%
Total sugars = 50 + 9 = 59%

Sucrose (as invert sugar) = $59 \cdot 5 - 50 \cdot 0 = 9 \cdot 5 \%$
From equation (2.2), Sucrose = $9 \cdot 5 \times 0 \cdot 95 = 9 \cdot 0 \%$
Total sugars = $50 \cdot 0 + 9 \cdot 0 = 59 \cdot 0 \%$

As the water content of the sample was found to be $29 \cdot 5 \%$, the soluble solids can be calculated as follows:

$$\text{Soluble solids } (\%) = \frac{\text{Total sugars } (\%)}{\text{Total sugars } (\%) + \text{Water } (\%)} \times 100$$

$$= \frac{59 \cdot 0}{59 \cdot 0 + 29 \cdot 5} \times 100 = 67 \%$$

Polarimetric methods

As sugars are optically active, sugar solutions possess the property of rotating the plane of polarisation of polarised light. The extent of the rotation varies according to the particular sugar, its concentration, the temperature and the length of tube used. The direction of rotation also varies with the sugar. The standard of rotation is represented by the specific rotation, $[\alpha]_D^{20}$, where 20 represents 20 °C and D denotes the use of sodium light of wavelength 589 nm. In general,

$$[\alpha]_D^{20} = \frac{100\alpha}{lc} \tag{2.3}$$

where α is the angular rotation of a solution of the pure sugar; l is the polarimeter tube length in decimetres; and c is the concentration of sugar in the solution.

The specific rotations at 20 °C of some of the commoner sugars are given in *Table 2.6*.

Table 2.6. SPECIFIC ROTATIONS OF THE COMMONER SUGARS AT 20 °C

Sugar	$[\alpha]_D^{20}$
Sucrose	$+66 \cdot 5$
Invert sugar	$-20 \cdot 0$
Dextrose	$+52 \cdot 7$
Laevulose	$-92 \cdot 7$
Lactose hydrate	$+52 \cdot 6$
Maltose	$+138 \cdot 4$

Note: Some sugars have high temperature coefficients, e.g., invert sugar has a zero rotation at 87 °C as the positive rotation of dextrose is then equal to the negative rotation of laevulose.

Mutarotation

A freshly prepared solution of a sugar has a different rotation from that attained on standing. The rotation of sugar solutions should be measured after equilibrium between the isomerides has been attained. This is achieved rapidly by adding a few drops of ammonia.

Determination of a single sugar in solution

If a solution is prepared containing $c\%$ w/v of a sample, the amount of a Sugar A present is given by

$$\text{Sugar A in sample } (\%) = \frac{\alpha}{[\alpha]_D^{20} \text{ for Sugar A}} \times \frac{100}{l} \times \frac{100}{c}$$

where α is the angular rotation of solution and l is the length of tube (dm).

Determination of sucrose

If sucrose is the only optically active substance in solution, its concentration can be estimated directly from the direct rotation using the equation above, which reduces to

$$\text{Sucrose } (\% \text{ w/v}) = \frac{100\,\alpha}{66 \cdot 5\,l}$$

If other optically active substances are also present, however, it is necessary to use the Clerget–Herzfeld method of double polarisation. On hydrolysis, dextrorotatory sucrose yields a mixture of equal parts of dextrose and laevulose (invert sugar), which is laevorotatory. The change of rotation is taken as the measure of the sucrose originally present.

METHOD 2.8B. DETERMINATION OF SUCROSE BY DOUBLE
POLARISATION

Pipette 50 ml of the prepared solution (concentration $X\%$ w/v) into a 100-ml volumetric flask and add 25 ml of water and 7 ml of 5 N hydrochloric acid. Immerse the flask in a water-bath at 60 °C for 12 min, mixing for the first 3 min. Then cool, dilute the solution to 100 ml at 20 °C and measure the rotation in a 2-dm tube (α_I). Also measure the rotation of the original solution (α_D), so that

$$[\alpha_D] = \frac{100\alpha_D}{lX} = \frac{100\alpha_D}{2X}$$

$$[\alpha_I] = \frac{100\alpha_I}{l\,X/2} = \frac{100\alpha_I}{2\,X/2} = \frac{100\alpha_I}{X}$$

Then, Sucrose (%) = $S = \dfrac{[\alpha_D]-[\alpha_I]}{0\cdot8865}$

The Inversion Divisor Factor (Q), $0\cdot8865$, corresponds to a change in rotation due to the inversion from $+66\cdot5°$ to $-22\cdot15°$, so that $+66\cdot5-(-22\cdot15) = 88\cdot65$. In practice, however, the actual value of Q varies according to the clearing agent used[22] and conditions of inversion (acid concentration, temperature, or whether invertase is used).

Calculation of concentration of two sugars, one of which is sucrose

From *Method 2.8b*, the percentage of sucrose (S) is first obtained. The proportion of the direct rotation due to sucrose can then be calculated and that due to the second sugar, X, obtained by difference, $[\alpha_D]-(0\cdot665 \times S)$, so that

$$X(\%) = \frac{[\alpha_D]-(0\cdot665 \times S)100}{[\alpha]_D^{20} \text{ for the second sugar}}$$

2.9. ACIDITY

TOTAL TITRATABLE ACIDITY

Although the pH value is the important factor in relation to preservation and the setting of certain gels, the total acid content of a food material is one of the simpler checks for controlling recipes. Acidic raw materials also have to be checked for purity. Many of the common acids used in the food industry (e.g., citric acid and tartaric acid) are the subject of monographs in the BP and BPC, which usefully describe the simplest titration method in each case.

The total titratable acidity (TTA) is most frequently determined by titrating with 0·1 N or 0·5 N sodium hydroxide using phenol-phthalein (pH 8·3–10·0) or bromothymol blue (pH 6·0–7·6) as indicator. Detection of the exact end-point is sometimes difficult, however, owing to the presence of buffers or dark colours in the food. Although in such instances some idea of the end-point can be deduced by using extra amounts of indicator and water for dilution, it is preferable to carry out the titration potentiometrically.

The TTA is usually reported in terms of the predominant acid present, e.g., milk as lactic acid, most fruits as citric acid, apple as malic acid and vinegar as acetic acid. In industrial control, it is not uncommon to report the acidity as 'No. of ml of 0·1 N alkali required by 10 g', because the titrations themselves are quite satisfactory when comparisons are the important factor. In some instances the acidity is reported in terms of the equivalence of the weight of an appropriate alkali. Acid phosphates intended for use in baking powder, for instance, are usually reported in terms of sodium bicarbonate (p. 229).

VOLATILE ACIDITY

The main volatile acid present in foods is acetic acid. With foods such as pickles (p. 261) the volatile acidity (VA) can be estimated by difference by titrating the acidity before (TTA) and after evaporating the sample down several times with water so that only the fixed acid (FA) remains (with all acidities expressed in terms of percentage of acetic acid):

$$\%VA = \%TTA - \%FA$$

The volatile acid can also be titrated in the distillate from the sample. Steam distillation, for which several special assemblies have been described[23], is preferable to direct distillation. The author has found with vinegar that the semi-micro distillation apparatus designed for Kjeldahl determinations (see *Method 2.6b*) is more efficient than macro assemblies. Accurate determination of the volatile acid is difficult, however, with both evaporation and distillation methods. For instance, only a fraction of the partially volatile acids such as lactic acid tend to volatilise. Also, some pyrolysis may occur, particularly during evaporation, causing breakdown of carbohydrates. Another possibility is coagulation of acid-binding colloids. Corrections can be applied for lactic acid or sulphur dioxide if necessary.

In routine control, the volatile acidity value is used as a measure of the acetic acid present. With alcoholic drinks, for instance, 'over-fermentation' gives rise to an objectionable taste. The proportion of acetic acid present in vinegar, pickles, sauces and mincemeat is also of considerable importance. For products such as pickles and salad cream, it is usual to calculate the percentage of volatile acid in the aqueous phase ($Avaq$):

$$Avaq = \frac{VA}{VA + W} \times 100$$

where VA is the percentage of volatile acid in the product and W is the percentage of water in the product; ($VA + W$) is virtually equivalent to the total volatile matter lost on evaporation. For a product depending on acetic acid for preservation, $Avaq$ should be at least 3·5%. Lower minima may be appropriate, however, if the product is heat processed or if salt and/or sugar is present (see also p. 261).

2.10. pH VALUE

The pH value is the common logarithm of the number of litres of solution that contains one gram-equivalent of hydrogen ion. Thus,

$$pH = -\log_{10}[H^+]$$

The pH scale ranges from 0 to 14, a value below 7 representing an acidic solution, 7 a neutral solution and above 7 an alkaline solution. The pH values of some common foods and materials are: citric acid 2·0, lemon 2·2–2·4, vinegar 2·4–3·4, apple 2·9–3·3, jam 3·0–3·5, tap water 4·5–8·5, bread 5·0–6·0, cabbage 5·2–5·4, flour 6·0–6·5, fish 6·0–6·3, milk 6·4–6·8, human blood plasma 7·3–7·5, and ammonia solution 11·5.

The pH can be assessed colorimetrically using suitable indicators, but is more accurately determined electrically. Most pH meters measure the potential difference between a standard calomel electrode and a glass electrode by balancing on a potentiometer. Suitable buffers for calibrating pH meters are as follows:

Buffer solution of pH 4 (0·0496 M potassium hydrogen phthalate): 10·12 g of $HKC_8H_4O_4$ (dried at 105 °C) per litre. This gives a pH of 4·002 at 20 °C and 4·008 at 25 °C.

Buffer solution of pH 9 (0·00997 M borax): 3·80 g of $Na_2B_4O_7$. $10H_2O$ per litre. This gives a pH of 9·22 at 20 °C and 9·18 at 25 °C.

The acidity as measured by the pH value is an important controlling factor in many natural and manufacturing processes. It is an especially important consideration in food preservation and storage, because of the inhibitive effect of acid on the growth of micro-organisms and enzymes. In general, bacteria are more sensitive to hydrogen ions than are yeasts and moulds. For most organisms there is a minimum and a maximum limit for growth and an optimum range for the most rapid growth. Hydrogen ions also influence the degree of heat applied in canning to achieve commercial sterility, e.g., vegetables and meat are processed at higher temperatures and for longer times than the more acid fruits.

The pH value also affects various physical properties of some foods, e.g., texture and setting power or gel strength of gelatine gels and the pectin–sugar–acid gel of jam.

Next to moisture, pH is probably the determination most frequently carried out in the food industry.

2.11. OXIDATION–REDUCTION POTENTIAL

The oxidation–reduction (redox) potential is the potential set up between a platinum electrode and the test solution and is a measure of the oxidising or reducing properties of the solution. Both oxidised and reduced forms must be present for a measurement to be made. Assuming that the pH is constant, the e.m.f. (E_h) at 25 °C of a cell consisting of a platinum electrode dipping into a solution containing both forms of the substance and connected to a standard hydrogen electrode is given by:

$$E_h = E_0 = \frac{0\cdot0591}{n} \log_{10} \frac{[Ox]}{[Red]}$$

where n is the number of electrons involved in the redox system; [Ox] is the molar concentration of the oxidised form; [Red] is the molar concentration of the reduced form; and E_0 is the standard oxidation–reduction potential of the system.

If [Ox] > [Red], then $E_h > E_0$, and if [Ox] = [Red], then $E_h = E_0$. The more positive the potential the greater is the oxidising power.

The method for determining oxidation–reduction potentials varies to some extent with the pH meter used, but it should not be difficult to make any necessary modifications from the following.

METHOD 2.11. DETERMINATION OF OXIDATION–REDUCTION POTENTIAL

Use the calomel electrode but connect a platinum electrode to the pH meter in place of the usual glass electrode and follow the instructions for reading in millivolts. Check the instrument against (*i*) pH 4·0 phthalate buffer (0·05 M potassium hydrogen phthalate) into which a little quinhydrone has been stirred (reading *a*) and (*ii*) 0·1 N hydrochloric acid also with quinhydrone added (reading *b*). The standard potential of the calomel electrode containing saturated KCl is 0·248 V at 20 °C (0·244 V at 25 °C). Assuming that readings are converted to volts at 20 °C, the following values should be obtained:

$$a + 0·248 = 0·471 \text{ V}$$

$$b + 0·248 = 0·642 \text{ V}$$

Determine the redox potential of the test solution (E_h) in the same way as for (*i*) and (*ii*) (reading $= X$ V), so that:

$$E_h = X + 0·248$$

If X is negative, reverse the electrode and from this reading $E_h = -X + 0·248$ V. If a 0·1 M KCl calomel electrode is used, the standard potential is 0·336 V both at 20 °C and 25 °C.

Apart from it being the basis of oxidation–reduction titrations, measurements of E_h are useful for correlating bacterial systems with the corresponding chemical reactions. The commonest applications are therefore in the fermentation, dairy, baking and meat industries.

In biochemistry, the exponent rH, which varies with pH, is sometimes used:

$$rH = 2\left(\frac{E_h}{0·0591} + pH\right)$$

The rH varies from 0 (hydrogen electrode) to 41 (oxygen electrode). In practice, the values found vary from 0 to 30.

2.12. ALCOHOL

Ethanol is normally determined by distilling a measured volume of sample and, after making up the distillate to the same volume, the alcohol content is assessed from its gravity by reference to appropriate tables. When the proportion of alcohol is relatively high, water is added to the sample before distilling it and the distillate is made up to twice the volume (double bulk) or four times the volume (quadruple bulk), for which appropriate tables giving the alcohol content (as percentages of alcohol and 'Proof Spirit') have been issued by H. M. Customs and Excise and published by H.M.S.O. Proof spirit (100%) has a sp. gr. of 0·91702 at 20/20 °C and contains 49·276% of ethanol by weight and 57·155% by volume.

The BP method for determining alcohol allows for the presence of various types of interfering substances. Thus Method I involves distilling into quadruple bulk and if the specific gravity and refractive index of the distillate correspond to the same amount of alcohol being present it is assumed that interference is minimal. In Method II, interference due to volatile compounds is prevented by saturating the diluted sample with salt and shaking it with light petroleum prior to distillation. Interference due to acidic volatile compounds can be prevented by making the solution alkaline before distilling it. Method III is used with products that contain oils, etc., which are liable to give emulsions when they are shaken with light petroleum. The extraction is then applied to the distillate obtained as in Method I.

If the amount of solid matter is small, as with spirits, the alcohol content can be obtained by determining the specific gravity of the sample directly and then making a correction for the alcohol 'obscured' by the dissolved solids. Each 1 % w/v of extract increases the specific gravity by 0·0041[24].

METHOD 2.12. DETERMINATION OF ALCOHOL BY THE BP METHOD

I. Measure 25 ml of sample into a 25-ml volumetric flask at 20 °C and wash it into a distilling flask (500–800 ml) with 100–150 ml of water. Add pumice, connect the flask to a still-head and condenser and distil at least 90 ml into a 100-ml volumetric flask. Dilute the distillate to the mark with water at 20 °C. Determine both the specific gravity (by using a specific gravity bottle or a pycnometer) and refractive index of the distillate at 20 °C. By reference to *Table 2.7* (quadruple bulk), if the refractive index differs by not more than 0·00007 (= 0·2 on the immersion refractometer scale), read off the percentage of ethanol corresponding to the specific gravity. If the difference exceeds 0·00007, treat 75 ml of distillate with sodium chloride and light petroleum (boiling range 40–60 °C) as in Method II, distil about 70 ml and dilute the distillate to 75 ml. If the refractive index and specific gravity values now correspond, assess the alcohol content from the specific gravity; otherwise some impurities are still likely to be present. If the distillate is turbid, proceed as in Method II. If steam-volatile acids are present, make the liquid just alkaline with 1 N sodium hydroxide, after adding *solid* phenolphthalein just prior to the final distillation.

II. Measure 25 ml of sample or distillate into a volumetric flask at 20 °C and wash it into a separator with 100 ml of water. Saturate the mixture with sodium chloride and shake it with 100 ml of light petroleum (boiling range 40–60 °C) for 3 min. Allow it to stand for at least 15 min and run the lower layer into a distillation flask. Shake

Table 2.7. ETHANOL (QUADRUPLE BULK) TABLE. (By courtesy of The Pharmaceutical Press.)

Specific gravity at 20°/20° of the distillate obtained by distillation to quadruple bulk	Percentage v/v of ethanol in the original preparation at 20°	Refractive index at 20° of the distillate obtained by distillation to quadruple bulk	Immersion refractometer readings*
0·9710	95·93	1·34661	50·3
0·9720	92·32	1·34605	48·8
0·9730	88·66	1·34549	47·3
0·9740	84·96	1·34493	45·8
0·9750	81·26	1·34437	44·3
0·9760	77·53	1·34380	42·8
0·9770	73·82	1·34324	41·3
0·9780	70·10	1·34267	39·8
0·9790	66·38	1·34211	38·3
0·9800	62·72	1·34154	36·8
0·9810	59·09	1·34098	35·3
0·9820	55·48	1·34044	33·9
0·9830	51·94	1·33991	32·5
0·9840	48·45	1·33942	31·2
0·9850	45·02	1·33892	29·9
0·9860	41·62	1·33842	28·6
0·9870	38·28	1·33796	27·4
0·9880	34·99	1·33751	26·2
0·9890	31·76	1·33705	25·0
0·9900	28·62	1·33663	23·9
0·9910	25·53	1·33620	22·8
0·9920	22·49	1·33578	21·7
0·9930	19·50	1·33540	20·7
0·9940	16·59	1·33501	19·7
0·9950	13·71	1·33466	18·8
0·9960	10·89	1·33432	17·9
0·9970	8·10	1·33397	17·0
0·9980	5·38	1·33362	16·1
0·9990	2·68	1·33331	15·3
1·0000	0·00	1·33300	14·5

* The readings refer to the scale proposed by Pulfrich and are applicable only to instruments calibrated in units corresponding to 14·5 = 1·33300; 50·0 = 1·34650; and 100 = 1·36464.

the light petroleum extract in the separator with 25 ml of saturated sodium chloride solution and, after allowing the mixture to separate, add the lower layer to the first brine solution run off. In certain instances, the BP requires a double separation to be carried out, in which case run the brine solution into a second separator and shake it with a further 100 ml of light petroleum before transference to the distillation flask. Then wash this second volume of light petroleum extract with the wash liquor from the first washing. Make the mixed solutions just alkaline with 1 N sodium hydroxide after adding *solid* phenolphthalein, add pumice and 100 ml of water, distil 90 ml into a 100-ml volumetric flask, make up to the mark with water at 20 °C and determine the specific gravity and refractive index as in Method I.

REFERENCES

1. NELSON, O. A. and HULETT, G. A., *Ind. Engng Chem., analyt. Edn.*, **12**, 40 (1920)
2. TATE, F. G. H. and WARREN, L. A., *Analyst, Lond.*, **61**, 367 (1936)
3. *Dean and Stark Apparatus*, BS 756: 1952, British Standards Institution, London
4. GEARY, P. G., *Control*, **7**, 303 (1963)
5. ANON., *Lab. Pract.*, **14**, 1313 (1965)
6. MITCHELL, J. and SMITH, D. M., *Aquametry*, Interscience, New York (1948)
7. *Determination of Water by the Karl Fischer Method*, BS 2511:1954, British Standards Institution, London
8. SANDELL, D., *J. Sci. Fd Agric.*, **11**, 671 (1960)
9. SOCIETY OF PUBLIC ANALYSTS, *Analyst, Lond.*, **68**, 276 (1943)
10. FILL, M. A. and STOCK, J. T., *Analyst, Lond.*, **69**, 121 (1944)
11. HARTLEY, A. W., *Analyst, Lond.*, **77**, 53 (1952)
12. BRADSTREET, R. B., *The Kjeldahl Method for Organic Nitrogen*, Academic Press, New York and London (1965)
13. MARKHAM, R., *Biochem. J.*, **36**, 790 (1942)
14. *Nitrogen Determination Apparatus (Micro-Kjeldahl)*, BS 1428:Part B1:1964, British Standards Institution, London
15. HOSKINS, J. L., *Analyst, Lond.*, **69**, 271 (1944)
16. YUEN, S. H. and POLLARD, A. G., *J. Sci. Fd Agric.*, **4**, 490 (1953)
17. CONWAY, E. J., *Microdiffusion Analysis and Volumetric Error*, 5th edn, Crosby Lockwood, London (1962)
18. TAYLOR, W. H., *Analyst, Lond.*, **82**, 488 (1957)
19. DOLBY, R. M., *J. Dairy Res.*, **28**, 43 (1961)
20. MACARA, T., *Analyst, Lond.*, **56**, 394 (1931)
21. LANE, J. H. and EYNON, L., *J. Soc. chem. Ind., Lond.*, **42**, 32T (1923)
22. SOCIETY OF PUBLIC ANALYSTS, *Analyst, Lond.*, **55**, 111 (1930)
23. JOSLYN, M. A., *Methods in Food Analysis: Physical, Chemical and Instrumental Methods of Analysis*, 2nd edn, Academic Press, New York, 420 (1970)
24. BEE, H. M., *J. Assoc. Publ. Analysts*, **8**, 97 (1970)

3

FOOD ADDITIVES

Preservatives—Antioxidants—Colouring Matters

The term 'additives' relates to substances that are deliberately added to foods by the processor. They are added to improve the keeping qualities of the food, or the appearance, texture or other organoleptic properties. Their addition is controlled by regulations, which usually prescribe those substances which can be present in foods on safety grounds as a result of feeding trials on animals[1]. The additives considered in this chapter are the main ones used at the time of publication.

PRESERVATIVES

The addition of preservatives to food is controlled by The Preservatives in Food Regulations 1962. The legal definition of 'preservative' excludes the classical substances such as salt, sugar, vinegar and alcohol. The main provision in the regulations is that several foods are permitted to contain named preservatives up to various prescribed limits. The commonest ones used are sulphur dioxide and benzoic acid, and details of their estimation are given below. Methods for determining nitrate and nitrite are given in Chapter 7.

SULPHUR DIOXIDE

Sulphur dioxide is the most commonly used food preservative. Although only a proportion of the preservative (the undissociated fraction of the unbound sulphurous acid) is effective against micro-organisms[2], the maxima stated in the regulations relate to the total amount present (calculated as SO_2 w/w). The total amount is usually determined by distilling the sample with acid and absorbing the sulphur dioxide with excess of an oxidising agent, which converts it into sulphuric acid. The lengthy reference method of Monier-Williams[3] using hydrogen peroxide has been made suitable for routine use by Shipton[4]. A rapid method quite commonly used is that involving direct titration with iodine as the sulphur dioxide distils over[5]. With liquids such as fruit juices and wines or with solids that can readily be dispersed in water, it is possible to release the bound sulphite with sodium hydroxide and, after acidifying, to titrate the total sulphur dioxide with iodine[6,7], thus avoiding the distillation process.

METHOD 3.1A. DETERMINATION OF SULPHUR DIOXIDE USING SHIPTON'S MODIFICATION[4] OF MONIER-WILLIAMS' REFERENCE METHOD

The main modifications from the earlier method are the use of (*a*) a less cumbersome vertical reflux condenser with ground-glass joints (*Figure 3.1*); (*b*) the replacement of gas by electrical heating; (*c*) the use of a regulated flow of nitrogen as inert gas because carbon dioxide may interfere with the end-point; (*d*) the abolition of the

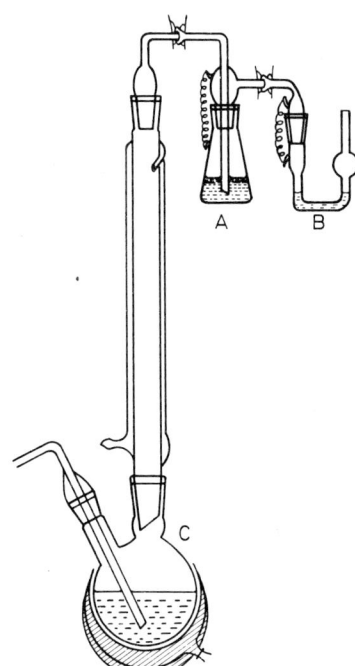

Figure 3.1. Shipton's apparatus for the determination of sulphur dioxide: The sample is boiled with acid in flask C and the sulphur dioxide produced is absorbed in hydrogen peroxide contained in flask A and guard-tube B

necessity of using previously boiled water; (e) the use of un-neutralised hydrogen peroxide, which can be corrected for by means of a blank titration; and (f) the omission of the heating of the condensers at the end of the refluxing.

Special reagents

Bromophenol blue Dissolve 0·4 g in 6 ml of 0·1 N sodium hydroxide and dilute the mixture with water to 100 ml.
0.05 N sodium hydroxide solution Standardise using bromophenol blue as indicator.

Procedure

Connect up the apparatus (*Figure 3.1*), ensuring that all joints are greased and tight when clamped. Surround flask C with a circular electric heating mantle. Remove flask A and the U-tube B and place in each 15 ml and 5 ml, respectively, of 3 % hydrogen peroxide. Add water to each (if necessary) so that bubbles pass through. To C add the prepared sample (*a*), 350 ml of water and 20 ml of concentrated HCl. Immediately, connect up the inert gas supply and adjust the rate of flow to 6–12 bubbles per minute in the U-tube. Switch on the electric heater to full heat. When the liquid boils, turn down the heat to give a slow, steady boil. Continue the boiling for 30 min, ensuring that the rate of flow of gas is maintained. Then disconnect the conical flask and the U-tube and transfer the solution in the latter to the former, washing it in with a little water. Add 3 drops of bromophenol blue indicator to the combined liquids in the flask and titrate with 0·05 N NaOH. Titrate the free acid in a further 20 ml of the hydrogen peroxide solution and subtract from the sample titration.

$$SO_2 \text{ (ppm)} = \frac{(\text{Sample titration} - \text{Blank titration}) \times 1600}{\text{Weight of sample taken (g)}}$$

Check on recovery of procedure

Prepare an approximately 1 % solution of sodium sulphite ($Na_2SO_3 \cdot 7H_2O$) Pipette 10 ml of 0·1 N iodine solution into a conical flask, add 1 ml of 5 N HCl and 40 ml of water. Run in slowly from a pipette 10 ml of the approximately 1 % sodium sulphite solution, swirling the flask throughout the addition to ensure completion of the reaction. Then titrate the excess of iodine with 0·1 N sodium thiosulphate, adding starch solution as indicator towards the end. Calculate the weight of sulphur dioxide (as SO_2) contained in the 10 ml of solution taken.

$$1 \text{ ml of } 0·1 \text{ N iodine} \equiv 0·003203 \text{ g of } SO_2$$

Then add to the distillation apparatus 10 ml of the sodium sulphite solution, 350 ml of water and 20 ml of concentrated HCl and carry out the procedure above. From the titration with standard NaOH, calculate the weight of sulphur dioxide which has reacted with the hydrogen peroxide and compare the result with the weight obtained from the iodine titration.

Note

   (a) Solid samples should be ground or minced. The amount to be used should preferably contain 3–15 mg of SO_2, but it is not usually desirable to use more than 50 g, particularly if frothing is expected.

METHOD 3.1B. DETERMINATION OF SULPHUR DIOXIDE BY TITRATION OF THE DISTILLATE WITH IODINE[5].

Apparatus (Figure 3.2)

The distillation apparatus (*Figure 3.2*) consists of a 500-ml round-bottomed flask fitted with a tap funnel and a twin-bulb still-head. The bung in the flask should preferably have an additional hole to make provision for steam distillation of some samples, see *Note (e)*. For general use, the hole can be closed by inserting a tight-fitting piece of glass rod.

The jacket of the long vertical condenser should be at least 45 cm long and the condenser tube 12 mm in diameter. The bottom of the adapter has twin bell-shaped portions containing holes to scrub the rapidly emerging gas. The bottom of the adapter should be about 8 mm from the bottom of the 600-ml beaker, which acts as the receiver. It is useful to have a stirrer in the beaker, preferably with a large flat, circular end so that it runs clear of the bell-shaped adapter. A 50-ml burette is clamped close and parallel to the condenser so that 0·05 N iodine solution can be added to the receiving beaker at any time during the titration.

81

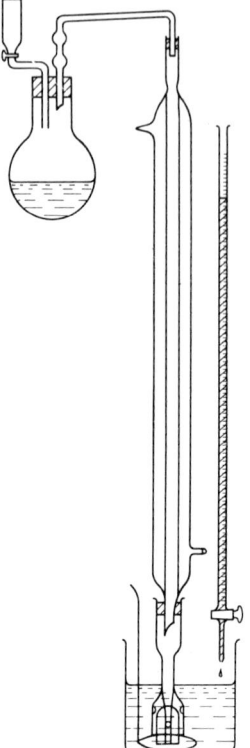

Figure 3.2. Apparatus for determining sulphur dioxide by using the 'Committee' method: The sample is boiled with acid and the sulphur dioxide is titrated directly with iodine as it distils over

Procedure

Place sufficient recently boiled and cooled distilled water in the beaker to cover the outer bell of the adapter and add 0·25 ml of 1 % starch solution. Add 0·05 N iodine dropwise from the burette with stirring until a pale blue colour is produced (*a*). Pour 200 ml of recently boiled and cooled distilled water into the distilling flask, followed by 25–100 g of sample and some glass beads. Connect up the apparatus. Run 25 ml of 20 % phosphoric acid solution (*b*) into the flask via the tap funnel and quickly close the tap. Heat the flask with the naked flame of a powerful burner so that the liquid boils in not more than $2\frac{1}{2}$ min. Add 0·05 N iodine from the burette into the receiving beaker (*c*) so that the pale blue colour is maintained throughout the titration (*d*). Continue the titration until the colour due to 0·1 ml of 0·05 N iodine persists for at least 1 min, which should be reached within a boiling time of 10 min (*e, f*).

1 ml of 0·05 N iodine ≡ 0·0016 g of sulphur dioxide

Example of calculation

50 g of sample required 8·3 ml of 0·05 N iodine.

$$\text{Sulphur dioxide in sample} = \frac{8·3 \times 0·0016 \times 10^6}{50}$$

$$= 266 \text{ ppm (as SO}_2\text{)}$$

Notes

(*a*) This should also be the colour at the end-point of the titration.

(*b*) For some materials, including gelatine and starch, concentrated hydrochloric acid should be used instead of phosphoric acid.

(*c*) The temperature of the distillate should be kept below 27 °C throughout the distillation.

(*d*) Almost all the sulphur dioxide comes over in the first rush of gas and only traces after 5 min of boiling. The addition of too great an excess of iodine may result in loss of iodine.

(*e*) Too prolonged boiling may produce reducing substances other than sulphur dioxide. If more than 10 min of distillation seems to be necessary to achieve the end-point requirements, the determination should be repeated using steam distillation. Superheated steam (not above 150 °C) may be advantageous with materials which give rise to excessive frothing (see original paper).

(*f*) The method is unsuitable for (*i*) foods containing other sulphur compounds, e.g., mustard and decomposed meat, and (*ii*) certain spices, e.g., nutmeg and mace. In (*i*), interference due to hydrogen sulphide from decomposed protein can be prevented by adding a small amount of copper acetate to the distillation flask prior to the distillation. When any doubt exists, however, the sulphur dioxide should be submitted to reflux distillation into hydrogen peroxide (*Method 3.1a*).

METHOD 3.1C. DETERMINATION OF SULPHUR DIOXIDE BY DIRECT TITRATION WITH IODINE

The fact that bound sulphurous acid can be released by alkali and the total sulphur dioxide then titrated with iodine after acidification was originally applied to wines by Ripper[2]. The method can also be applied to other liquids and to solid foods that can be dispersed in water. Therefore, although the following method was devised by Potter[7] for the determination of sulphur dioxide in dehydrated cabbage, it can readily be applied to many other foods.

Procedure

In each of two 600-ml beakers (A and B), suspend 8 g of ground dehydrated cabbage in 400 ml of water and add 5 ml of 5 N NaOH. Stir the mixture gently so that air is not beaten into the solution, and allow it to stand for 20 min. Then to A add 7 ml of 5 N HCl and 10 ml of 1% starch solution and titrate the solution rapidly with 0·05 N iodine to give a measure of the total iodine-reducing power.

To determine the reducing material other than sulphite in B, treat the mixture with alkali and acid as for A, but immediately after the acidification add 2 ml of 3% hydrogen peroxide to oxidise sulphite to sulphate. Then add starch and titrate the solution with 0·05 N iodine to give a measure of the reducing power other than that due to sulphite.

Assuming that an 8-g sample was used:

$$\text{Sulphur dioxide (ppm)} = (\text{Titre A} - \text{Titre B}) \times 200$$

Joslyn and Braverman[2] have also discussed the estimation of the free sulphur dioxide by direct titration with iodine (see also p. 268).

BENZOIC ACID

Although benzoic acid is usually incorporated into foods as its more soluble sodium salt for the purpose of compliance with regulations, the amount is calculated as the acid (w/w). For many foods, benzoic acid can be extracted from the acidified material with an organic solvent. If emulsions are formed that are difficult to eradicate, however, it is necessary to steam distil the acid collecting the distillate in excess of alkali. Other acids that may have distilled over or are extracted can be destroyed by treating the warmed alkaline distillate with potassium permanganate solution.

METHOD 3.2A. DETERMINATION OF BENZOIC ACID AFTER SEPARATION
BY DIRECT EXTRACTION

Mark a 500-ml volumetric flask at 400-ml with a grease pencil.
Weigh out 100 g of sample into a beaker and transfer it to the 500-ml
volumetric flask with water (*a*). Add 10 ml of 10 % NaOH and 120 g
of NaCl and adjust the volume to 400 ml. Allow the flask to stand
for at least 1 h with frequent shaking. Then make up the solution to
the 500-ml mark, mix and filter. Pipette 100 ml of the filtrate into a
separator, neutralise it with 3 N HCl to litmus paper and add 4 ml
of acid in excess. Extract the benzoic acid with 50 ml of chloroform
(*b*). Shake the mixture and allow it to stand for 30 min. Run off the
separated lower layer (*c*) and filter. Pipette 25 ml of filtrate into a
flask and evaporate off the chloroform gently (*d*). Dissolve the
residue in 50 ml of 80 % alcohol and titrate either with 0·05 N NaOH
using phenolphthalein as indicator or potentiometrically to pH 8·2
(*a* ml). Also titrate 50 ml of the 80 % alcohol (*b* ml) as a blank.

$$1 \text{ ml of } 0.05 \text{ N NaOH} \equiv 0.0061 \text{ g of benzoic acid}$$

Notes

(*a*) Solid samples should first be macerated in a mechanical
blender.

(*b*) The hardness of shaking varies according to the product. If
difficulties arise owing to the formation of emulsions, repeat
with another 100 ml of filtrate and shake less vigorously.

(*c*) If an emulsion that is difficult to break forms, as with foods
of syrupy consistency, such as sauces, the method should be
abandoned and the distillation process (*Method 3.2b*) used
instead.

(*d*) Benzoic acid is the only major preservative acid that is not
destroyed by permanganate. Therefore, if there is a possibility that other acids are present, it is advisable to remove
the acid from the chloroform extract by shaking with
NaOH. Then other acids can be destroyed by adding
permanganate to the alkaline solution at 45 °C (see *Method
3.2b*) and, after acidifying, the benzoic acid can be re-
extracted.

METHOD 3.2B. DETERMINATION OF BENZOIC ACID AFTER SEPARATION
BY STEAM DISTILLATION

To a 500-ml steam distillation flask add 30–100 g of sample and

200 ml of water, saturate the solution with salt (40 g per 100 ml) and make it distinctly acid with phosphoric acid. Rapidly steam distil (a) 500 ml into a large porcelain basin containing 10 ml of 1 N NaOH. Wash down the condenser with 25 ml 0·1 N NaOH and evaporate the distillate down to about 25 ml on a water-bath. Cool the solution to about 45 °C and add 5% potassium permanganate solution until a pink colour persists after stirring (b). Then decolorise the solution with SO_2 solution and add sufficient dilute sulphuric acid to dissolve the precipitated manganese dioxide and make the liquid acidic. Transfer the solution to a graduated separator, saturate it with salt (33 g per 100 ml) and extract the benzoic acid four times with 15-ml volumes of diethyl ether (c). Wash the combined ether extracts twice with a small volume of water and then filter into a small flask, washing through with a little more ether. Evaporate off the ether (preferably at room temperature) and dissolve the residue of benzoic acid in the flask in 2 ml of acetone, add 2 ml of water and titrate with 0·05 N NaOH using phenol red as indicator (d).

1 ml of 0·05 N NaOH ≡ 0·0061 g of benzoic acid

Notes

(a) Heat the flask throughout the steam distillation.

(b) The permanganate destroys volatile acids other than benzoic acid. If cinnamon is present, however, the cinnamic acid is partially converted into benzoic acid during the oxidation.

(c) In the reference method of Monier-Williams[8], the warmed ether is removed in a test-tube by aspiration. The residue so obtained is then heated at 160 °C so that the benzoic acid crystals sublime on to the sides of the tube and are then weighed.

(d) Prepare the phenol red indicator as follows: warm 50 mg of phenol red with 2·85 ml of 0·05 N NaOH and 5 ml of alcohol (90%), and when it has dissolved, make up to 250 ml with 20% alcohol.

OTHER PERMITTED PRESERVATIVES

Methods for the determination of other permitted preservatives, e.g., p-hydroxybenzoates[9], nitrites and nitrates[10] (see p. 99), propionic acid[11,12], sorbic acid[13], and diphenyl and o-phenyl-phenol[14], have also been described.

ANTIOXIDANTS

Antioxidants are added to oils and fats in order to delay the onset of oxidative rancidity. Such additions are controlled by The Antioxidant in Food Regulations 1966, in which the main provisions are that certain oils and fatty materials may contain various gallates, butylated hydroxyanisole (BHA) and butylated hydroxytoluene (BHT).

GENERAL SCHEME FOR THE EXAMINATION OF OILS AND FATS FOR GALLATES, BUTYLATED HYDROXYANISOLE AND BUTYLATED HYDROXYTOLUENE

The examination of oils and fats for the presence of antioxidants normally involves shaking out the additive with ethanol or methanol and applying the appropriate reactions to the separated extract. A useful general scheme in which qualitative and quantitative tests can be applied to a common methanolic extract has been published by the APA[15].

METHOD 3.3A. EXTRACTION OF ANTIOXIDANT

Weigh out 10 g of oil or melted fat into a Mojonnier tube (*Figure 3.3*), shake it vigorously with 25 ml of 95 % methanol and place it in water at about 45 °C. When the layers have separated reasonably well, pour most of the upper layer into a 50-ml volumetric flask.

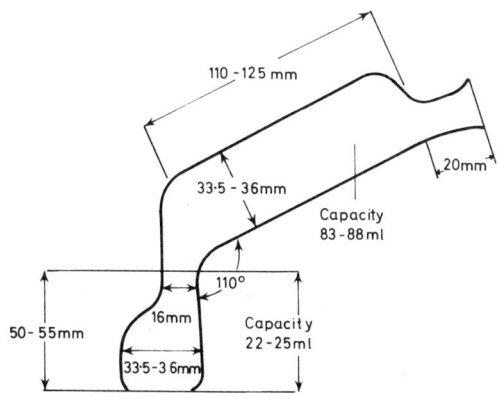

Figure 3.3. Mojonnier extraction tube. (After BS 1741, 1963, p. 20)

Then add 20 ml of methanol to the tube, re-extract the mixture with methanol, add the upper layer to the 50-ml flask and dilute to the mark with more methanol. Add 1 g of calcium carbonate to the flask and shake and filter the mixture, rejecting the first runnings. The filtrate can be used for the identification and determination of all the five principal permitted antioxidants. The extraction of BHT is not complete, however, amounting only to about half of that present in the oil.

METHOD 3.3B. EXAMINATION OF THE METHANOL EXTRACT FOR GALLATES[16]

Shake 10 ml of methanol extract, 1 ml of acetone and 10 mg of powdered ammonium ferrous sulphate for 1 min, and, after allowing the mixture to stand for 30 min, measure the optical density, d, of the blue colour against 10 ml of extract plus 1 ml of acetone in a 1-cm cell at 580 nm.

$$\text{Gallate (mg per 2-g sample)} = dk$$

where k = 0·622 for n-propyl gallate
     = 0·785 for n-octyl gallate
     = 0·952 for n-dodecyl gallate

Chromatographic methods for the identification of individual gallates are given in the APA Report[15].

METHOD 3.3C. EXAMINATION OF THE METHANOL EXTRACT FOR BUTYLATED HYDROXYANISOLE (BHA)

Gibbs reagent 0·01% 2,6-dichloro-*p*-benzoquinone-4-chloroimine in 95% methanol.

Procedure

To a 25-ml glass-stoppered, graduated cylinder add 2 ml of methanol extract (by pipette) and then 2 ml of methanol (95%) and 8 ml of 0·5% aqueous borax solution. Add 2 ml of Gibbs reagent by pipette, and, after allowing the mixture to stand for 15 min, make it up to 20 ml with n-butanol. Mix and measure the optical density (A) of the blue colour against water at 610 nm in a 1-cm cell. At the same time, measure the optical density of a reagent blank (B) using methanol instead of the methanol extract from the sample and that of a BHA standard containing 250 μg of 95% methanol per 10 ml (C). The BHA concentration is given by:

$$\frac{A-B}{C-B} \times 50 \ \mu g \text{ BHA in the 2 ml of methanol extract taken}$$

METHOD 3.3D. EXAMINATION OF THE METHANOL EXTRACT FOR BUTYLATED HYDROXYTOLUENE (BHT)

Dianisidine reagent Dissolve 250 mg of *o*-dianisidine (3,3'-dimethoxybenzidine) in 50 ml of methanol, shake the mixture for 5 min with 100 mg of activated charcoal and filter. Mix 40 ml of clear filtrate with 60 ml of 1 N HCl. Prepare the reagent solution fresh as required and protect it from light.

Procedure

To a 100-ml separator (*a*) add 5 ml of methanol extract (by pipette), 5 ml of water and 2 ml of dianisidine reagent solution. Stopper the separator, mix the contents carefully, add 0·8 ml of sodium nitrite solution (0·3%) and mix thoroughly. After allowing the mixture to stand for 5 min, add 4 ml of chloroform and extract the BHT complex by shaking vigorously for 30 s. Allow the mixture to stand for a few minutes to effect separation of the layers. Then place 0·8 ml of pure methanol in a 5- or 10-ml measuring cylinder (*a*), add the lower chloroform layer in the separator up to the 4-ml mark and mix well. Rapidly measure the optical density of the purplish red solution at 520 nm in a 1-cm cell against a mixture of 2 ml of methanol plus 8 ml of chloroform. Compare the value obtained with those from the reference standards and multiply the result by two (*b*).

Reference standards

Prepare a solution containing 50 mg of BHT per 100 ml of methanol. Dilute this stock solution 100-fold with 50% v/v methanol to give a concentration of 5 μg/ml. Then prepare further dilutions (again using 50% methanol) to cover the range 1–5 μg/ml as the reference standards. In each instance take 10 ml of reference solution and 2 ml of dianisidine reagent (*a*) and continue as for the sample commencing with the sentence 'Stopper the separator, mix the contents carefully, ...'

Notes

> (*a*) All glassware should be amber coloured, or protected from light by covering it with metal foil.
>
> (*b*) This multiplication is necessary as only about half of the BHT present is extracted originally from the sample by the 95% methanol used. Doubtful results should be checked by separating the antioxidant following steam distillation[17].

COLOURING MATTERS

The addition of colouring matters to foods is controlled by The Colouring Matter in Food Regulations 1966 (as amended), which gives a list of compounds which may be used. Those most commonly used are the permitted water-soluble coal-tar colours covered in *Table 3.1*. It should be borne in mind, however, that it is not permissible to add colour to many foods, such as some dairy products, tea and coffee, and unprocessed flesh foods, fruits, and vegetables.

The water-soluble coal-tar colours are usually detected by one or more of the following methods:

(*a*) paper chromatography;
(*b*) thin-layer chromatography;
(*c*) chemical tests;
(*d*) spectrophotometry.

In all instances comparisons must be made with the known colours rather than relying on published data, which may have been obtained under different conditions.

Solutions of solid colours as received as raw materials can usually be chromatographed, etc., directly, but with foodstuffs the colour has to be isolated as in *Method 3.4d*, in which white wool is used for the removal and purification of the dye.

METHOD 3.4A. IDENTIFICATION OF WATER-SOLUBLE ARTIFICIAL COLOURS USING PAPER CHROMATOGRAPHY

Solvents

1.	Ammonia, sp. gr. 0·880	1 volume
	Water	99 volumes
2.	Aqueous sodium chloride solution (2·5% w/v)	
3.	Sodium chloride solution (2% w/v in 50% ethanol)	
4.	Isobutanol	10 volumes
	Ethanol	20 volumes
	Water	10 volumes

Permitted colour	1	2	3	4	5	6	7	8	9	10	11	12
REDS—												
Ponceau 4R	0·87	0·39	0·51	0·32	0·18	0·26	0·13	0·26	0·25	0·07	0·57	0·33
Carmoisine	0·48	0·08	0·59	0·54	0·44	0·17	0·37	0·28	0·55	0·30	0·15	0·05
Amaranth	0·62	0·15	0·27	0·29	0·14 (0·46)	0·19 (0·49)	0·11	0·17	0·16	0·04	0·33	0·07
Red 10B	0·51	0·14	0·36	0·37	0·26	0·30	0·23	0·37	0·37	0·21	0·20	0·00, 0·14
Erythrosine BS	0·21	0·03	0·52	0·61	1·00 (0·00)	0·58	0·47	0·57	0·52, 0·36–1·00	0·56, 0·20–0·68	0·06	0·00
Red 2G	0·70	0·18	0·42	0·43	0·34	0·35	0·38	0·39	0·41	0·18	0·46	0·16
Red 6B	0·48	0·08	0·20	0·26	0·18	0·17	0·37	0·22	0·22	0·10	0·28	0·05
Red FB	0·11	0·01	0·10	0·22	0·25 (0·08)	0·11	0·49	0·13	0·00–0·58	0·24, 0·00–0·35	0·01	0·00
Fast Red E	0·44	0·10	0·60	0·50	0·38	0·47	0·45	0·49	0·51	0·24	0·19	0·06
ORANGES—												
Orange G	0·91	0·61	0·78	0·51	0·35	0·47	0·48	0·52	0·52	0·23	0·66	0·54
Orange RN	0·25	0·08	0·84	0·80 (0·44)	0·59 (0·35)	0·75	0·74 (0·32)	0·75	0·78	0·57 (0·19)	0·28	0·09
Sunset Yellow FCF	0·73	0·29	0·72	0·46	0·28	0·45	0·40	0·43	0·46 (0·81)	0·22	0·43	0·24
YELLOWS—												
Tartrazine	0·91	0·32	0·41	0·28	0·12	0·17	0·09	0·20	0·25	0·04	0·70	0·33
Yellow 2G	0·96	0·81	1·00	0·63	0·44	0·41	0·41	0·37	0·65	0·31	0·76	0·75
GREENS, BLUES AND VIOLETS—												
Green S	0·96	0·70	1·00	0·63	0·44	0·44	0·70	0·41	0·67	0·30	0·83	0·90
Indigo Carmine	0·52	0·12	0·27	0·25	0·14	0·20	0·30	0·28	0·34	0·14	0·15	0·13
Violet BNP	0·83	0·00–0·88	1·00	0·72 (0·42)	0·54	0·63 (0·30–1·00)	0·80 (0·00–1·00)	0·68 (0·31, 0·20–1·00)	0·75 (0·25–1·00)	0·32	0·00–0·94	0·89
BROWNS, BLACKS—												
Brown FK	0·08, 0·26	0·03 (0·16)	0·00–0·78	0·39 (0·55)	0·18 (0·00–0·49)	0·39	0·36, 0·49, 0·84 (0·19, 0·62)	0·75, 0·57 (0·31)	0·77, 0·61 (0·33)	0·49, 0·27 (0·06)	0·18, 0·03	0·00, 0·12
Chocolate Brown FB	0·00–1·00	0·00–1·00	0·00–1·00	0·22	0·00–0·49	0·13	0·00–0·50	0·15 (0·00–0·54)	0·15 (0·00–0·80)	0·00 (0·00–0·30)	0·00–1·00	0·00–1·00*
Chocolate Brown HT	0·00–1·00	0·00–1·00	0·00–1·00	0·22	0·00–0·47	0·13	0·00–0·50	0·15 (0·00–0·54)	0·15 (0·00–0·80)	0·00 (0·00–0·30)	0·00–1·00	0·00–1·00*
Black PN	0·00–0·80	0·06	0·10	0·17	0·05	0·06	0·10	0·13 (0·00–0·56)	0·14 (0·56)	0·02 (0·26)	0·10	0·02 (0·14)
Black 7984	0·00–0·68	0·05	0·06	0·12	0·07	0·10	0·03	0·11	0·00	0·03	0·06	0·00

* Position of leading edge of streak varies with concentration.

5. n-Butanol   40 volumes

5. n-Butanol	40 volumes
Glacial acetic acid	10 volumes
Water	24 volumes
6. Iso-butanol	30 volumes
Ethanol	20 volumes
Water	20 volumes

Then to 99 volumes of this mixture add 1 volume of ammonia solution of sp. gr. 0·880.

7. Phenol	80 g
Water	20 g
8. Ethyl methyl ketone	350 volumes
Acetone	150 volumes
Water	150 volumes
Ammonia, sp. gr. 0·880	1 volume
9. Ethyl methyl ketone	70 volumes
Acetone	30 volumes
Water	30 volumes
10. Ethyl acetate	11 volumes
Pyridine	5 volumes
Water	4 volumes

11. Dilute 5 ml of ammonia solution (sp. gr. 0·880) with water to 100 ml and dissolve 2 g of trisodium citrate in the solution

12. Water	30 volumes
Hydrochloric acid, sp. gr. 1·18	6·5 volumes

Procedure

Cut a sheet of Whatman No. 1 chromatography paper so that it forms a cylinder within a 1-litre beaker (or of another suitable size). Open out the paper and draw a line parallel to the bottom and about 2 cm away from it. Place a spot of the concentrated solution of the unknown dye on the line together with a series of spots (2 cm apart) of aqueous solutions of known dyes of similar colour and dry them. Roll the paper into a cylinder, staple the ends together (without them quite touching) with an office stapling machine and place it in the beaker containing a 1-cm layer of solvent. The solvent to be used can be selected by reference to *Table 3.1*. For example, Solvent 1 is better able to differentiate between the permitted red colours than Solvent 2. Cover the beaker with a clock-glass immediately after inserting the paper. After the solvent front has moved up to a distance of about 2 cm from the top, remove the paper. Then mark the solvent front with a pencil, dry the paper in a current of hot air (e.g., from a hair dryer) and compare known and unknown spots (*Figure 3.4*). By referring again to *Table 3.1*, run further chromatograms using

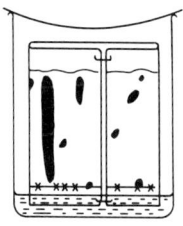

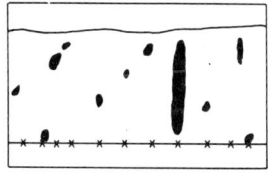

Figure 3.4. Paper chromatography of colours: The upper diagram shows the separation of the dyes on the paper contained in a beaker. The lower diagram shows the separated spots on the paper after running and drying the chromatogram

different solvents. If more than one colour appears to be present, cut out the separate spots, dissolve them in aqueous acetone, evaporate the solution to dryness, re-dissolve the residue in 2 or 3 drops of water and run further chromatograms.

$$R_\text{F} \text{ value} = \frac{\text{Distance moved by dye spot}}{\text{Distance moved by solvent}}$$

METHOD 3.4B. IDENTIFICATION OF WATER-SOLUBLE ARTIFICIAL COLOURS USING THIN-LAYER CHROMATOGRAPHY

Solvents

1. Phenol	80 g	
Water	20 g	
2. Isopropanol	100 volumes	
Ammonia, sp. gr. 0·880	10 volumes	
Water	10 volumes	
3. Aqueous potassium nitrate solution (saturated)		

4. Acetoacetic ester   50 volumes
     Methanol   20 volumes
     Ammonia, sp. gr. 0·880   10 volumes
5. Acetoacetic ester   50 volumes
     Pyridine   20 volumes
     Ammonia, sp. gr. 0·880   10 volumes
6. Ethyl acetate   70 volumes
     Pyridine   30 volumes
     Water   10 volumes

Preparation of plates

Prepare a slurry containing Kieselgel G with twice its weight of water and spread it on to 20 × 10-cm glass plates to give thin layers of 250 μm thickness. Heat the plates at 160 °C for 1–2 h and cool and store them in a desiccator containing silica gel.

Procedure

Apply 5–10 μl of concentrated aqueous solutions of the colours by pipette at a distance of about 2 cm from the edge of the plate and at least 1·5 cm apart. Place the spotted plate in a suitable enclosed tank containing the appropriate solvent—most colours can be separated using Solvents 2 and 3 (Dickes[18]). Sufficient time should previously be given for the attainment of equilibrium between the solvent and the atmosphere within the tank. After the solvent front has moved about 12 cm, remove the plate from the tank, dry it in a stream of hot air (a domestic hair dryer is adequate for the purpose) and compare the spots. If it is considered necessary to run the colour from a spot again (as when a mixture is present), extract the spot with aqueous acetone, centrifuge or filter the extract and, after concentrating it, apply it to further chromatographic runs.

METHOD 3.4C. IDENTIFICATION OF WATER-SOLUBLE ARTIFICIAL COLOURS BY SPECTROPHOTOMETRY

Dilute the pure neutral dye solution to a suitable intensity of colour and obtain the absorption curve (preferably on a recording spectrophotometer) in neutral (0·02 % ammonium acetate), acid (0·1 N HCl) and alkaline solution (0·1 N NaOH). Compare the maxima with those obtained with the pure known dye[19–21].

METHOD 3.4D. EXTRACTION OF SYNTHETIC WATER-SOLUBLE COLOURS
FROM FOODSTUFFS

Principle

All the permitted water-soluble coal-tar colours are acidic and will
dye wool in acidic solution. The colour is therefore first obtained in
acidic solution and then boiled with wool and purified. The concentrated solution is then used for chromatography, spectrophotometry
or chemical tests. Starch, fat, etc., in the food may interfere and have
first to be removed by suitable means.

Preparation of wool Boil *pure* white knitting wool in *dilute* ammonia
and then in water.

Extraction from acid foods To an acidic liquid food such as a soft
drink, add a few drops of dilute acetic acid and boil the mixture with
the prepared wool. With sweets, jam, etc., dissolve them in a small
volume of water before adding the acetic acid.

Extraction from starchy foods With custard powder, flour confectionery, etc., grind the sample with 2% of ammonia in 70%
alcohol, allow the mixture to stand and centrifuge or filter it.
Evaporate the extract on a water-bath, stir the residue with water
and dilute acetic acid, boil it with wool and proceed as for solutions
of acidic foods.

Extraction from fatty foods Grind meat products (or other fatty
foods) thoroughly with 50% alcohol or acetone and a small volume
of ammonia solution of sp. gr. 0·880. Allow the mixture to stand for
at least 30 min, then filter it. Evaporate the filtrate on a water-bath
and extract the residue with water, add acetic acid and wool and
proceed as for solutions of acidic foods. If much fat is present,
remove it by shaking with light petroleum.

Notes on other aspects of the examination of food colours

(a) R_F values of numerous non-permitted colours (including
basic ones) have been reported[22, 23].

(b) A preliminary check on whether mixtures of different
colours may be present can be made by drying a large spot
on a filter-paper. Then, on adding a drop of water, separate
streaks can often be seen if two quite different colours are
present, such as blue and yellow.

(c) All the permitted water-soluble colours are acidic and dye
wool in acidic solution. Basic colours can be isolated from

foodstuffs by boiling with wool in dilute ammoniacal solution.

(d) Some colours (especially basic ones) fluoresce strongly. Therefore, it is often useful to examine and compare chromatographic spots under ultraviolet light as confirmatory evidence.

(e) *Oil-soluble colours* can be separated by reverse-phase paper chromatography[24]. The absorption peaks for the permitted oil-soluble colours in chloroform are Oil Yellow GG at 380 nm and Oil Yellow XP at 405 nm.

REFERENCES

1. PEARSON, D., *Rev. Nutr. Fd Sci.*, No. 17, 16 (1969)
2. JOSLYN, M. A. and BRAVERMAN, J. B. S., *Adv. Fd Res.*, **5**, 97 (1954)
3. MONIER-WILLIAMS, G. W., *Analyst, Lond.*, **52**, 343; 415 (1927)
4. SHIPTON, J., *Fd Preserv. Q.*, **14**, 54 (1954)
5. PRESERVATIVES DETERMINATION COMMITTEE, *Analyst, Lond.*, **53**, 118 (1928)
6. JENSEN, H. R., *Analyst, Lond.*, **53**, 133 (1928)
7. POTTER, E. F., *Fd Technol., Champaign*, **8**, 269 (1954)
8. MONIER-WILLIAMS, G. W., *Analyst, Lond.*, **52**, 153; 229 (1927)
9. THACKRAY, G. B. and HEWLETT, A., *J. Assoc. Publ. Analysts*, **2**, 13 (1964)
10. ADRIAANSE, A. and ROBBERS, J. E., *J. Sci. Fd Agric.*, **20**, 321 (1969)
11. WALKER, G. H., GREEN, M. S. and FENN, C. E., *J. Assoc. Publ. Analysts*, **2**, 2 (1964)
12. WALKER, G. H. and GREEN, M. S., *J. Assoc. Publ. Analysts*, **3**, 87 (1965)
13. CARR, W. and SMITH, G. A., *J. Assoc. Publ. Analysts*, **2**, 37 (1964)
14. GUNTHER, F. A., BLINN, R. C. and BARKLEY, J. H., *Analyst, Lond.*, **88**, 36 (1963)
15. ASSOCIATION OF PUBLIC ANALYSTS, *The Detection and Determination of Antioxidants in Food*, Special Report No. 1, Association of Public Analysts, London (1963)
16. CASSIDY, W. and FISHER, A. J., *Analyst, Lond.*, **85**, 295 (1960)
17. SZALKOWSKI, C. R. and GARBER, J. B., *J. agric. Fd Chem.*, **10**, No. 6, 490 (1962)
18. DICKES, G. J., *J. Assoc. Publ. Analysts*, **3**, 49 (1965)
19. PEARSON, D., *The Chemical Analysis of Foods*, 6th edn, Churchill, London, 60–61 (1970)
20. ASSOCIATION OF PUBLIC ANALYSTS, *Separation and Identification of Food Colours Permitted by the Colouring Matters in Food Regulations, 1957*, Association of Public Analysts, London (1960)
21. PEARSON, D., *J. Assoc. Publ. Analysts*, **5**, 37 (1967)
22. PEARSON, D. and CHAUDHRI, A. B., *J. Assoc. Publ. Analysts*, **2**, 22 (1964)
23. PEARSON, D., *J. Assoc. Publ. Analysts*, **4**, 61 (1966)
24. SILK, R. S., *J. Ass. off. agric. Chem.*, **42**, 427 (1959)

4

TRACE ELEMENTS

In food analysis, the term 'trace elements' is often used to refer to those elements (mainly metals) which are present in relatively small amounts in foods and for which limits have been officially recommended or prescribed, as listed in *Table 4.1*.

Table 4.1. STATUTORY AND RECOMMENDED GENERAL LIMITS FOR TRACE ELEMENTS (ppm)

Element	General statutory limit	General recommended limit (FSC)
Arsenic	1(a)	—
Copper	—	20(a)
Lead	2(a)	—
Tin	—	250(b)
Zinc	—	50
Fluorine	30(c)	—

Notes
  (a) There are special limits in the case of several named food-stuffs. Lower limits apply, for instance, to liquids such as drinks and higher limits to dried foods such as spices and tea.
  (b) Applies to canned foods.
  (c) Calculated on the acid phosphate present. The limit for baking powder is 15 ppm.

Such limits are applied in most of these instances because of (i) possible toxicity of the element, and (ii) the feasibility of the limit in relation to good manufacturing practice. It should be borne in mind, however, that although some elements (e.g., F, Cu, Zn) are harmful if taken in large amounts, small amounts are necessary for normal metabolism[1]. This Chapter, however, also includes methods for the

97

determination of some nutritive non-toxic elements, such as calcium, iron and phosphorus.

THE ESTIMATION OF TRACE ELEMENTS

In most instances, the estimation of trace elements in foods involves the following steps:
  (i) Production of an acidic solution of the inorganic elements in the food after removal of the organic matter by dry ashing or wet oxidation[2];
 (ii) Removal or masking of interfering substances;
(iii) Determination of the selected element, usually by using colorimetry.

It is especially important when determining trace elements to clean all apparatus thoroughly, to use the purest grades of reagents and to carry out blank determinations.

Results of trace element determinations are expressed as parts per million (ppm), usually on a w/w basis.

DESTRUCTION OF ORGANIC MATTER

METHOD 4.1A. DESTRUCTION OF ORGANIC MATTER BY DRY ASHING

Clean a non-etched silica dish with concentrated HCl/concentrated HNO_3/water mixture and then with water alone by boiling the liquid in the dish in each case. Then heat the dish in a hot bunsen burner flame, cool it and weigh into it (by using a semi-accurate balance) a suitable amount of sample (containing, say, 2–5 g of dry solids). Heat the dish gently on a tripod in a fume cupboard (liquid and semi-solid samples should first be partially dried on a water-bath or hot-plate). When smoke is no longer evolved, transfer the dish into a muffle furnace and complete the ashing at a temperature preferably not exceeding 450 °C (appreciable losses by volatilisation occur above this temperature with some elements). Then take up the ash in the appropriate acid (see methods for individual elements below).

METHOD 4.1B. DESTRUCTION OF ORGANIC MATTER BY WET OXIDATION

Using a semi-accurate balance, transfer a suitable amount of sample (containing, say, 2–5 g of dry solids) into a 200–650-ml Kjeldahl digestion flask and add 20 ml of concentrated nitric acid and 10–20 ml of water, depending on the amount of water in the sample. Boil the mixture for 10 min until the volume is about 20 ml, cool it, and add 10 ml of concentrated sulphuric acid. Boil the mixture and add concentrated nitric acid *immediately* the liquid begins to blacken (*a*). Continue the heating and immediate addition of more concentrated nitric acid in small volumes until the liquid no longer blackens. *Continue the heating for a considerable period* until *copious* white fumes are produced. Then cool the solution (*carefully!*), add 5–10 ml of saturated ammonium oxalate solution (*b*) and boil it until copious white fumes are again produced. Cool the solution before diluting it with water (*c*).

Notes

(*a*) Delay may cause loss of the elements being determined.

(*b*) This assists in breaking down residual nitric acid, the presence of which seriously affects subsequent procedures.

(*c*) If more than one element has to be determined on the same sample, make up the diluted digest to 50 ml with water and use appropriate aliquots, e.g., 25 ml for Pb, 15 ml for As, 10 ml for Cu.

A further method of wet oxidation for metals which tend to be lost in the above technique, such as mercury, is described on p. 208.

ARSENIC

METHOD 4.2. DETERMINATION OF ARSENIC USING SILVER DIETHYLDITHIOCARBAMATE[3, 4]

Principle of method

The solution from the wet oxidation of the sample is transferred to a generator flask and treated with potassium iodide and stannous chloride to convert arsenic to the trivalent state. Zinc is added so that the liberated hydrogen generates arsine, which is allowed to bubble through a solution of silver diethyldithiocarbamate in pyridine, yielding a pink coloration which is measured spectrometrically at 540 nm.

The large amounts of arsenic sometimes found in shellfish such as crab are probably in the non-toxic organic form. If the sample is placed into the reaction flask (without wet oxidation) and treated directly with zinc and acid, the arsine produced is derived solely from toxic inorganic arsenic.

Apparatus

A suitable apparatus is shown in *Figure 4.1*. It consists of a 100-ml conical flask, A, with a connecting tube, B, containing lead acetate

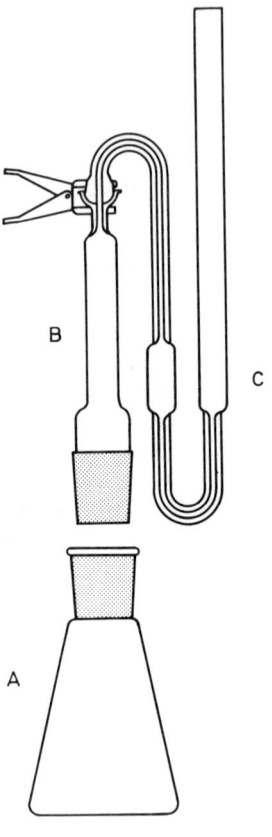

Figure 4.1. *Arsenic apparatus permitting absorption of arsine in silver diethyldithiocarbamate solution:* A, 100-ml flask; B, connecting tube to trap H_2S; C, absorption tube. (From *Fisons Review*, April 1966, No. 22, p. 10)

wool to trap any hydrogen sulphide produced. This tube is connected by means of a spring clip to the absorption tube, C, into which is placed the carbamate reagent.

Special reagents

N.B. It is most important to use 'arsenic-free' reagents. Special grades are obtainable from some suppliers.

Stannous chloride solution Dissolve 40 g of stannous chloride dihydrate in a mixture of 25 ml of water and 75 ml of concentrated HCl.

Silver diethyldithiocarbamate 0.5% in water-white pyridine.

Lead acetate wool Saturate absorbent cotton-wool with 10% lead acetate trihydrate solution, drain it, press it tightly and dry it under vacuum.

Stock standard arsenic solution Dissolve 0.132 g of dried As_2O_3 in 20 ml of 35% NaOH solution. Slowly add, with stirring, 100 ml of water, then 10 ml of concentrated sulphuric acid and finally dilute the mixture with water to 1000 ml (1 ml of solution $\equiv 0.1$ mg of As).

Standard arsenic solution Dilute 5 ml stock arsenic solution to 500 ml with water immediately before using (1 ml of solution $\equiv 1$ μg of As).

Procedure

Substances (particularly chemicals) that are readily soluble in water or HCl can be placed directly into the conical flask A of the apparatus (*Figure 4.1*). Otherwise, wet-oxidise the sample (*Method 4.1b*) with sulphuric and nitric acids, making absolutely sure that every trace of nitric acid is removed before proceeding. Dilute the solution to 40 ml with water and transfer it to the conical flask of the apparatus (the 40 ml of solution in the flask should not contain more than 10 μg of As). Then add 10 ml of concentrated HCl, 2 ml of 15% KI solution and 2 ml of stannous chloride solution. Mix and allow to stand for 15 min. During this period, lightly pack the top third of the connecting tube B with lead acetate wool and assemble the apparatus. Add, by pipette, 5.0 ml of silver diethyldithiocarbamate solution to the absorption tube C. At the requisite time, disconnect the conical flask, add 5 g of granular zinc and rapidly re-assemble the apparatus. After the reaction has proceeded for 45 min, disconnect the absorption tube and tilt it to-and-fro to mix and dissolve any red solid. Then measure the optical density in a 1-cm cell against water at 540 nm. Compare the reading against the calibration curve, making due allowance for the blank. Antimony also produces a colour with an absorption peak at 515 nm. Arsenic can, however, be separated from interfering elements by distillation as the trichloride[5].

Preparation of calibration curve

To a series of 100-ml conical flasks (A) add 0, 2, 4, 6, 8 and 10-ml volumes of standard arsenic solution ($\equiv$ 0–10 μg of As). Then add sufficient water to give a total volume in each flask of 40 ml. To each flask add 10 ml of concentrated HCl, 2 ml of 15% KI solution and 2 ml of stannous chloride solution and continue as for the sample above, allowing the gas produced with zinc to bubble through 5 ml of reagent. Construct the calibration curve relating optical density at 540 nm to micrograms of As in each flask.

OTHER METHODS FOR DETERMINING ARSENIC

Arsenic can also be determined by using the Gutzeit[5, 6] or molybdenum-blue methods[5].

CALCIUM

METHOD 4.3. DETERMINATION OF CALCIUM AS THE OXALATE BY TITRATION WITH PERMANGANATE[7]

Principle of method

Calcium is precipitated at about pH 4 (to prevent interference by phosphate) as the oxalate, which is dissolved in sulphuric acid and the liberated oxalic acid is titrated with standard potassium permanganate solution.

Reagents

Dilute ammonia Mix one volume of ammonia solution of sp. gr. 0·880 with two volumes of water.
Dilute sulphuric acid Slowly add 20 ml of concentrated sulphuric acid to 180 ml of water with stirring, then cool the solution.
Dilute acetic acid Mix one volume of glacial acetic acid with two volumes of water.
Bromocresol green 0·05% in alcohol.

Procedure

Ash a suitable amount of sample (containing up to 50 mg of Ca) at 500 °C. Cool the ash, add 10 ml of 5 N HCl and evaporate the mixture on a water-bath. Then add 10 ml of 2 N HCl, bring the solution just to boiling, dilute it with with 10 ml of water and filter it into a 250-ml beaker. Repeat the acid extraction, dilution and filtration. Add bromocresol green to the filtrate and make it *just alkaline* with dilute ammonia. Acidify the solution with dilute acetic acid and then add 0·5 ml of glacial acetic acid. Heat the solution to boiling and slowly add 10 ml of saturated ammonium oxalate solution. Then add dilute ammonia to the hot solution until it becomes yellow–green (pH ∼ 3·8) and allow it to stand for at least 4 h (preferably overnight). Filter the solution through a Whatman No. 44 filter-paper, checking the first runnings for unprecipitated calcium by adding ammonium oxalate solution. Wash the filter with very dilute ammonia solution (1 volume of ammonia solution of sp. gr. 0·880 plus 50 volumes of water) until the filtrate no longer gives a precipitate on addition of dilute nitric acid and silver nitrate solution. Pierce the paper with a pointed glass rod and wash the precipitate into a beaker flask with a fine spray of cold water from a polythene wash-bottle, then with 10 ml of warm (60 °C) dilute sulphuric acid and finally with hot water to give a final volume of about 150 ml. Titrate the solution with 0·05 N potassium permanganate at 75 °C (stir with a thermometer) until a pink colour persists for 30 s.

$$1 \text{ ml of } 0·05 \text{ N } KMnO_4 \equiv 0·001002 \text{ g of Ca}$$
$$\equiv 0·001402 \text{ g of CaO}$$

OTHER METHODS FOR DETERMINING CALCIUM

Calcium can be titrated with EDTA, provided precautions are taken against interfering ions[8, 9]. Other techniques available are based on nephelometry[10, 11] and flame photometry[11].

COPPER

METHOD 4.4. DETERMINATION OF COPPER[12]

Principle of method

Citrate and EDTA are added to the copper solution, which is adjusted to pH 8·5. The solution is shaken with a solution of diethylammonium diethyldithiocarbamate in carbon tetrachloride. The brownish yellow copper complex formed is extracted from the aqueous layer into the carbon tetrachloride and the optical density is measured at 436 nm. For modifications when bismuth or tellurium is present, the original paper[12] should be consulted.

Reagents

Acid extractant Mix together carefully 20 ml of concentrated HCl, 10 ml of concentrated HNO_3 and 30 ml of water.

6 N *ammonia* If a special grade for foodstuff analysis is not available, purify the ammonia by shaking it with the carbamate solution (see below) until no more colour is removed.

Diethylammonium diethyldithiocarbamate solution 0.1% in carbon tetrachloride.

EDTA–citrate solution Dissolve 20 g of ammonium citrate and 5 g of disodium dihydrogen ethylenediaminetetra-acetate in water and dilute the solution to 100 ml. Purify it by shaking with 15-ml portions of carbamate solution (see above) until no more colour is removed.

Thymol blue Dissolve 0.1 g of thymol blue by warming it with 4.3 ml of 0.05 N NaOH + 5 ml of 90% v/v ethanol and dilute the solution to 250 ml with 20% v/v ethanol.

Standard copper (*stock solution*) 0.393 g of $CuSO_4 \cdot 5H_2O$ per litre of 2 N sulphuric acid (1 ml of solution $\equiv$ 100 μg of Cu).

Standard copper (*working solution*) Dilute 10 ml of stock solution as required with 2 N sulphuric acid to a total volume of 500 ml (1 ml of solution $\equiv$ 2 μg of Cu).

Procedure

Ash a suitable amount of sample (usually 2–10 g) at 600 °C in a silica dish. Boil the ash gently with 10 ml of acid extractant with constant stirring and wash the mixture into a separator with water (filtering if necessary) so that the total volume is 25 ml. Add 10 ml of EDTA–citrate solution, 5 drops of thymol blue and then 6 N ammonia until the solution is green or bluish green after cooling. Add 15 ml of diethylammonium diethyldithiocarbamate solution and shake the mixture vigorously for 2 min. Push a piece of cotton-wool up the stem of the separator, run the lower layer into a 1-cm spectrophotometer cell and measure the optical density without delay (to avoid fading) at 436 nm against the reagent blank. Calculate the amount of copper present in the sample by reference to the calibration graph.

Preparation of calibration graph

To a series of separators, add 10-ml volumes of EDTA–citrate

104

solution and the amounts of working standard copper solution and
2 N sulphuric acid given in *Table 4.2*. Shake each vigorously with
15 ml of diethylammonium diethyldithiocarbamate solution as for
the sample and construct the calibration graph relating optical
density at 436 nm to micrograms of copper.

Table 4.2. MIXTURES TO BE USED FOR THE PREPARATION OF THE CALIBRATION GRAPH
FOR COPPER

Working standard copper solution/ml	0	1	2·5	5	10	15	20	25
2 N sulphuric acid/ml	25	24	22·5	20	15	10	5	0

The working standard solution contains 2 μg ml^{-1} of Cu.

OTHER METHODS FOR DETERMINING COPPER

Various carbamates, all of which give yellow complexes with copper,
have been used for its determination. One of the commonest is
sodium diethyldithiocarbamate, which also gives a stable complex in
alkaline solution[13].

Precautions against interference are normally unnecessary if the
acid solution from the sample is shaken directly with a solution of
various salts of dibenzyldithiocarbamic acid[14]. Neocuproine appears
to be specific for copper[12].

IRON

METHOD 4.5. DETERMINATION OF IRON USING 2,2′ dipyridyl[15]

Reagents

Acetate buffer Dissolve 16·6 g of anhydrous sodium acetate (dried
at 100 °C in water, add 24 ml of glacial acetic acid and dilute the
mixture with water to 200 ml.
Hydroquinone solution Dissolve 2·5 g of hydroquinone in water, add
0·5 ml of concentrated HCl and dilute the mixture with water to
100 ml.
2,2′ Dipyridyl solution 0·1 % in water.
Standard iron solution Dissolve 3·512 g of $Fe(NH_4)_2(SO_4)_2 \cdot 6H_2O$
in water, add 2 drops of 5 N HCl and dilute the mixture with water
to 500 ml. Then, when required, dilute 10 ml of this solution to 1 litre
(1 ml of solution $\equiv$ 0·01 mg of Fe).

Procedure

Clean a silica dish by heating HCl in it and then wash it thoroughly with water, taking precautions against contamination with metal from the tongs used. Ash a suitable amount of sample (e.g., 2–10 g of flour) at 550 °C. Add 2 ml of concentrated HCl, cover the dish with a clock-glass and place it on a water-bath for 30 min. Carefully wash any drops of condensed water into the dish using a little water from a wash-bottle. Then transfer the ash solution down a rod, using several small volumes of water, into a 100-ml volumetric flask. Make up the volume to the mark with water, then mix and filter. To a boiling-tube add 10 ml of filtrate (or a smaller aliquot made up to 10 ml), 3 ml of acetate buffer, 2 ml of hydroquinone solution and 2 ml of 2,2'-dipyridyl solution. Mix and measure the optical density of the solution in a 1-cm cell against water at 520 nm. Also perform a blank and compare both readings with the calibration curve to obtain the concentration of iron in the original sample.

Preparation of calibration curve

To a series of boiling-tubes add 0, 0·5, 1·0, 1·5, 2·0, 3·0 and 4·0-ml volumes of the diluted standard iron solution (containing 0·01 mg ml^{-1} of Fe), dilute each to exactly 10 ml and add reagents as for the sample. Construct the calibration curve relating optical density at 520 nm to micrograms of Fe in each tube.

OTHER METHODS FOR DETERMINING IRON

Potassium thiocyanate and o-phenanthroline[16] also give red colorations with iron that are suitable for its determination. Both Fe(II) and Fe(III) are estimated with thioglycollic acid.

LEAD

METHOD 4.6. DETERMINATION OF LEAD USING SODIUM SULPHIDE[17]

Principle of method

The lead is dissolved from the ash with nitric acid. The iron present is converted into the thiocyanate, which is removed by extraction with ether–amyl alcohol. The solution is made alkaline and sodium sulphide is added to produce a brown colloidal suspension of lead sulphide, the colour of which is assessed against a Nessleriser disc. Colour due to copper is inhibited by the addition of cyanide. It is normally adequate as a routine sorting test, provided that the instructions are closely followed and the purest reagents are used.

Special reagents

Concentrated ammonium citrate solution Dissolve 200 g of citric acid in water, *gradually* stir in ammonia solution of sp. gr. 0·880 (approximately 170 ml) until the solution is alkaline to litmus, then cool and transfer it to a separator. Shake the solution successively with 10-ml volumes of 0·03 % diphenylthiocarbazone in chloroform until the lower layer remains green. Then wash the aqueous layer with 15-ml volumes of chloroform until the washings are colourless. Dilute the aqueous layer to 500 ml with water.
Dilute ammonium citrate solution Dilute 110 ml of concentrated ammonium citrate solution with water to 500 ml.
Dilute ammonia Mix 100 ml of ammonia solution of sp. gr. 0·880 and 200 ml of water.
EAA mixture Mix equal volumes of diethyl ether and amyl alcohol just prior to use.

Procedure

N.B. Rinse *all* glassware, e.g., separators and Nessler glasses, with *distilled* water just prior to use.

Ash 10 g of a solid food or 20 g of a semi-solid food (or 200 g of a liquid after evaporation on a water-bath) at not more than 450 °C in a silica dish. Cool the ash, add 3 ml of concentrated nitric acid and evaporate the mixture to dryness on a water-bath. Then add a further 3 ml of nitric acid, stirring the residue thoroughly on the water-bath with a glass rod for 2 min before stirring in 17 ml of water and heating the mixture for a further 1 min. Filter the mixture into separator A and wash the dish and paper with 10 ml of water. Cool and add 20 ml of EAA to the mixture, shake it, then add 5 ml of saturated ammonium thiocyanate and shake it again vigorously. Separate the mixture and run the lower aqueous layer into separator B. Then shake the remaining EAA layer in A with 10 ml of water and add this wash-water to B. Shake the bulked aqueous liquid in B with 10-ml volumes of EAA until the upper layer is no longer pink (run off the aqueous layer each time into separators that have previously been rinsed with distilled water). Run the washed aqueous layer into a 100-ml measuring cylinder and add 5 ml of 20% ammonium acetate, 5 ml of dilute ammonium citrate, 2 ml of 10% KCN and 10 ml of dilute ammonia. Dilute the mixture to 90 ml with water and mix. If the liquid is clear, make the volume up to 100 ml and mix. If the liquid is cloudy, add enough concentrated ammonium citrate to give a clear solution after mixing, make up the volume with water to 100 ml and mix.

Divide the 100 ml of solution equally into two 50-ml Nessleriser glasses and place one in the left-hand compartment of the Nessleriser. To the second glass add 3 drops of 10% sodium sulphide solution, mix and place the mixture in the right-hand compartment. Match the colour due to lead in half of the sample taken originally against the appropriate disc ($10\gamma \equiv 0{\cdot}01$ mg of Pb).

Example of calculation

10 g of sample taken.
After dividing the solution:

$$5\ \text{g} \equiv 30\gamma \equiv \frac{30}{5}\ \text{ppm} \equiv 6\ \text{ppm}$$

Reference procedure

In the SAC[18] reference procedure, which is more complicated and less suitable for routine use, dithizone is used for the elimination of some interfering substances and for the determination of the metal. The lead is finally extracted at pH 9–9·5 and the colour of the red dithizonate measured spectrophotometrically at 520 nm.

MERCURY

A general method of estimating mercury in food is given on p. 208.

PHOSPHORUS

METHOD 4.7A. DETERMINATION OF PHOSPHATE BY THE PHOSPHOMOLYBDATE VOLUMETRIC METHOD

Principle of method

Metaphosphates and pyrophosphates must first be converted to the orthophosphate, which is treated at 65–70 °C with excess of molybdate in the presence of nitric acid. The yellow precipitate of ammonium phosphomolybdate formed reacts quantitatively with standard alkali, the excess of which is titrated with acid.

Molybdate reagent Mix 75 ml of ammonia solution of sp. gr. 0·880 with 225 ml of water. Add this gradually, whilst shaking, to a 1-litre volumetric flask containing 125 g of molybdic acid plus 100 ml of water. When dissolution is complete, add 400 g of ammonium nitrate and dilute to the mark with water. Then add this solution to 1 litre of nitric acid (sp. gr. 1·19), mix, maintain the mixture at 37 °C for 24 h, then filter it.

Procedure

Prepare an acidic solution of the ash of the sample or use that obtained from wet oxidation. Then, to a solution containing about 0·02–0·04 g of P_2O_5 (*a*), *neutralised* with ammonia, add 5 ml of concentrated nitric acid per 100 ml (the total volume should preferably not exceed 200 ml). Warm the liquid to 65–70 °C and add 60 ml of the molybdate reagent. Stir the mixture and keep it at 65–70 °C for 15–20 min so that the precipitate settles out, then cool and filter it, preferably on a Gooch crucible. Wash the yellow precipitate with 1 % nitric acid, then with 1 % ammonium nitrate solution and finally with a small volume of water (*b*). Transfer the crucible containing asbestos and the precipitate to a beaker flask. Add some water and a measured slight excess of 0·5 N (*c*) NaOH (from a burette), using phenolphthalein as indicator. Mix well, making sure that the precipitate is completely dissolved *without warming* and back-titrate with 0·5 N HCl (*c*). The difference between the volumes of 0·5 N NaOH and 0·5 N acid represents the amount of phosphate present.

$$1 \text{ ml } 0\cdot5 \text{ N} \equiv 0\cdot001544 \text{ g of } P_2O_5$$

A different factor must be used if the liquid is boiled after addition of the standard alkali:

$$1 \text{ ml } 0\cdot5 \text{ N} \equiv 0\cdot001365 \text{ g of } P_2O_5$$

Notes

(*a*) The phosphorus must be in the form of the orthophosphate. This can be ensured by boiling the solution with nitric acid.

(*b*) The precipitate is slightly soluble in water. The triple washing procedure can be replaced by washing only with 1 % potassium nitrate.

(*c*) Smaller amounts can be titrated using 0·1 N solutions, but the precipitate then takes longer to dissolve.

METHOD 4.7B. DETERMINATION OF PHOSPHATE BY THE VANADIUM PHOSPHOMOLYBDATE COLORIMETRIC METHOD[19]

Principle of method

The method is based on Misson's reaction, in which the phosphorus present as the orthophosphate reacts with a vanadate–molybdate reagent to produce a yellow–orange complex, the optical density of which is measured at 420 nm (cf., Donald *et al.*[20]).

Special reagents

Vanadomolybdate reagent Dissolve separately 20 g of ammonium molybdate and 1 g of ammonium vanadate in water and mix the solutions. Acidify the mixture with 140 ml of concentrated nitric acid and dilute it with water to 1 litre.
Standard phosphate solution Dissolve 1·917 g of potassium dihydrogen phosphate (dried at 105 °C) in water and dilute the solution to 1 litre (stock solution). Dilute 50 ml to 250 ml (1 ml of solution $\equiv$ 0·2 mg of P_2O_5).

Procedure

Weigh out a suitable amount of sample (e.g., 5 g if solid; more if liquid) into a silica dish, add 1 g of calcium oxide and a small volume of water (unless the sample is a liquid). Mix, and after washing off any material on the stirring rod used, incinerate the mixture at 500 °C. Transfer the cooled ash to a beaker with the assistance of 20 ml of water and the careful addition of 12 ml of concentrated HCl and then 5 ml of concentrated nitric acid. Simmer for 15 min and filter into a 250-ml volumetric flask (if any appreciable amount of insoluble matter remains on the filter, ignite the paper, dissolve the ash in acid and add the solution to the filtrate). Make up the volume to the mark with water and mix. Add 25 ml of a dilution containing about 6 mg of P_2O_5 to a 100-ml volumetric flask, then add 25 ml of vanadomolybdate reagent and immediately dilute the mixture to 100 ml with water. At the same time, transfer 25 ml of standard phosphate solution (1 ml of solution $\equiv$ 0·2 mg of P_2O_5) into a second 100-ml volumetric flask and also add to it 25 ml of reagent and dilute the mixture. Measure the optical density of both at 420 nm in 1-cm cells 10 min after making up the volume to the mark. Estimate the phosphorus content of the sample from the difference in extinctions between the two yellow solutions and by reference to the calibration graph.

Preparation of calibration graph

Using a burette, add to a series of 100-ml volumetric flasks exactly 25, 26, 27, 28, 29, 30 and 31 ml of standard phosphate solution ($\equiv 5 \cdot 0$, $5 \cdot 2$, $5 \cdot 4$, $5 \cdot 6$, $5 \cdot 8$, $6 \cdot 0$ and $6 \cdot 2$ mg of P_2O_5). Add 25 ml of vanadomolybdate reagent to each solution and immediately dilute them to the marks with water. Mix and measure the apparent extinction of each solution ($\equiv 5 \cdot 2$–$6 \cdot 2$ mg of P_2O_5) in 1-cm cells at 420 nm, referred to the $5 \cdot 0$ mg P_2O_5 solution as standard. Hence plot the readings against milligrams of P_2O_5.

Note that the colorimetric reactions in the above method must be carried out as close as possible to 20 °C.

OTHER METHODS FOR DETERMINING PHOSPHATE

Phosphorus can also be determined by using the volumetric quinoline phosphomolybdate method of Wilson[21], the molydenum-blue colorimetric method of Tschopp and Tschopp[22] or the ascorbic acid procedure of Fogg and Wilkinson[23].

POTASSIUM

METHOD 4.8. DETERMINATION OF POTASSIUM BY FLAME PHOTOMETRY

Principle of method

Assuming that interfering substances such as salts of calcium, iron or aluminium are present, the ash is first boiled with ammonium oxalate solution. The potassium content is estimated from the measurement of the characteristic radiation at 766–770 nm, by using a suitable filter, from a flame into which a solution of the sample is sprayed.

Reagents

Ammonia solution Dilute 60 ml of ammonia solution of sp. gr. 0·880 with water to 200 ml.

Potassium stock solution Dissolve 5·779 g of potassium dihydrogen phosphate (dried at 105 °C) in water and dilute the solution to 1 litre.

Standard diluted potassium solution Dilute 50 ml of stock potassium solution to 1 litre with water ($\equiv 100$ ppm as K_2O). Prepare from the stock solution just before use.

Procedure

With solutions of fruit products, apply flame photometry directly to a filtered solution. With other samples containing organic matter, proceed as follows.

Incinerate a suitable amount of sample at not more than 500 °C in a silica dish. Wash the cooled ash into a 400-ml beaker with 125 ml of water, add 50 ml of saturated ammonium oxalate solution and boil the mixture for 30 min. Cool, add the ammonia solution in slight excess, wash into a 500-ml volumetric flask with water, make up the solution to the mark, then mix and filter it. Dilute the mixture so that the final solution contains about 15 ppm of K_2O (solution A).

Prepare a series of solutions from the standard diluted potassium solution containing 10, 12, 14, 16, 18 and 20 ppm of K_2O. Set the sensitivity of the flame photometer so that 100 scale divisions (full deflection) is equivalent to 20 ppm. Spray each standard solution at least twice, checking the sensitivity between each reading against the 20 ppm solution. From the readings obtained, prepare a calibration graph.

Reset the instrument at 100 scale divisions with the 20 ppm solution and spray the diluted sample solution A. Estimate the potassium content from the calibration graph after taking several readings.

OTHER METHODS FOR DETERMINING POTASSIUM

More conventionally, potassium can be determined gravimetrically due to the insolubility of the perchlorate or the chloroplatinate in alcohol[19].

TIN

METHOD 4.9A. DETERMINATION OF TIN BY TITRATION WITH IODINE

Principle of method

After wet oxidation of the sample, the tin is reduced to the stannous form (using aluminium plus hydrochloric acid), which is titrated with iodine solution. Interfering metals are seldom present in sufficient amounts in foods to affect the interpretation of the value obtained for tin. If necessary, however, the tin can be separated as the sulphide by passing hydrogen sulphide through the acidic solution. The volumetric method is normally adequate and convenient in routine control of canned foods, where the main consideration is compliance with the comparatively high limit of 250 ppm.

Procedure

Place a convenient amount of material (usually 10–25 g) in a large Kjeldahl digestion flask. Add 35 ml of concentrated nitric acid, mix and then carefully add 15 ml of concentrated sulphuric acid. Cool the solution if necessary to prevent frothing. Heat the solution on the digestion stand (very gently at first—beware frothing!) until no more brown fumes are evolved and the liquid begins to turn black. *Immediately* this happens, add more concentrated nitric acid and continue the heating and addition until a colourless solution is obtained. *Cool* the solution and carefully add 1 g of potassium chlorate and then 15 ml of concentrated hydrochloric acid. Heat the liquid until white fumes are evolved. Obtain a 200–300 ml conical flask fitted with a Contat–Gockel (or similar) valve (*Figure 4.2*) and have some 10% potassium bicarbonate solution ready. Remove the valve and transfer the liquid from the digestion flask, using successive small volumes of water, into the conical flask. Add to the conical flask 0·5 g of aluminium foil cut into small pieces and 25 ml of concentrated hydrochloric acid. Fit the flask immediately with the valve almost filled with the 10% potassium bicarbonate solution. Heat the flask until the solution boils and then allow it to cool, *adding more bicarbonate solution to the valve so that it is always nearly full*. After cooling, the solution should be quite clear, i.e., *all the aluminium should have dissolved*. Then remove the valve and add a few pieces of marble down the side of the flask and about 1 ml of 1% starch solution. Titrate the solution *as rapidly as possible* with

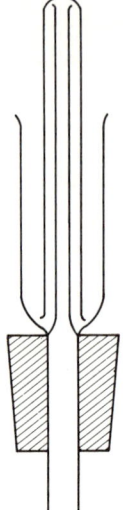

Figure 4.2. Snelling valve for retaining the solution in an inert atmosphere during reduction of tin

0·01 N iodine solution to the usual starch blue end-point. Carry out a blank determination.

1 ml of 0·01 N iodine ≡ 0·000595 g of Sn

Example of calculation

25 g taken
Titration = 2·2 ml of 0·01 N iodine
Blank   = 0·1 ml of 0·01 N iodine

$$(2\cdot2 - 0\cdot1) \times 0\cdot000595 \times \frac{10^6}{25} = 50 \text{ ppm of Sn}$$

The recommended maximum limit for canned foods is 250 ppm of tin.

METHOD 4.9B. COLORIMETRIC DETERMINATION OF TIN USING QUERCETIN[24]

Principle of method

The sample is wet oxidised. An alcoholic solution of quercetin, when added to the acidic solution of the sample containing tin (IV),

114

produces a yellow–orange colour the intensity of which is measured at 437 nm. Interference from iron(III) is masked by thiourea.

Special reagents

2,4-Dinitrophenol solution 0·1 % in 50 % aqueous ethanol.
Quercetin 0·2 % in ethanol.

Procedure

Weigh out a suitable amount of the well mixed sample (containing not more than 1000 μg of Sn) and transfer to a (preferably silica) digestion flask. Add 10 ml of concentrated nitric acid and, after 10 min, 5 ml of concentrated sulphuric acid. Boil the solution vigorously using a Meker burner, adding further *small* volumes of nitric acid immediately charring occurs. Continue boiling after charring has ceased until copious white fumes of SO_3 are produced. Cool the solution, add 20 ml of water, transfer to a 50-ml volumetric flask, make up to the mark and mix. Pipette 2·0 ml into another 50-ml volumetric flask and add 0·2 ml of 2,4-dinitrophenol solution as indicator. Add 10 % sodium carbonate solution dropwise until a yellow colour just appears. Discharge the colour by the dropwise addition, with shaking, of 2·5 N HCl and then add 5 ml in excess. Add 3 ml of saturated aqueous thiourea solution, 5 ml of quercetin reagent, 25 ml of ethanol and immediately dilute the solution to the mark with water and mix. Allow the mixture to stand for 30 min and measure the optical density in a 2- or 4-cm cell at 437 nm against a reagent blank. Compare the reading against the calibration curve.

Preparation of calibration curve

Dissolve 0·0500 g of pure tin in 50 ml of boiling concentrated sulphuric acid. Cool the solution and add it slowly to 120 ml of water, mixing continuously. Transfer the cooled solution to a 200-ml volumetric flask and dilute to the mark with 25 % sulphuric acid so that 1 ml of solution contains 250 μg of Sn. To a series of digestion flasks each containing 5 ml of concentrated sulphuric acid plus 10 ml of concentrated nitric acid, add 0·0, 1·0, 2·0, 3·0 and 4·0 ml of this standard tin solution. Boil each solution thoroughly to expel the nitric acid and until copious white fumes are produced. Cool the solutions, add 20 ml of water to each, transfer them to 50-ml volumetric flasks, make up to the mark and proceed as for the sample. Take all readings 30 min after making up to the mark. Construct a graph relating optical density to micrograms of tin.

OTHER METHODS FOR DETERMINING TIN

The volumetric and colorimetric methods for tin described above are suitable for the fairly large trace amounts for which canned foods are normally examined in routine work. The colorimetric determination of smaller amounts of tin using dithiol has been improved by the use of zinc dithiol[25] as reagent and sodium lauryl sulphate[26, 27] as dispersant.

ZINC

METHOD 4.10. DETERMINATION OF ZINC BY EXTRACTIVE TITRATION USING DITHIZONE[28]

Principle of method

A standardised chloroform solution of dithizone is added from a burette to an aqueous solution of zinc from the ash of the sample. After shaking the mixture thoroughly following each addition, the end-point of the titration is reached when no further coloured dithizonate is formed, indicating that all the zinc has reacted.

Special reagents

Dithizone extraction solution Extract 40 ml of a stock dithizone solution (0·1 % in chloroform) with 50 ml of 0·08 N ammonia and reject the lower layer. Acidify the ammoniacal extract with dilute hydrochloric acid and extract twice with 50-ml portions of chloroform. Reject the aqueous layer and dilute the extract with chloroform so that 1 ml of solution $\equiv$ 10 μg of Zn (approx.)—see standardisation method (p. 117).
Standard zinc solution Dilute 10 ml of stock solution (0·44% $ZnSO_4 \cdot 7H_2O$) to 1000 ml (1 ml of solution $\equiv$ 10 μg of Zn).
Bromothymol blue Warm 0·1 g of bromothymol blue with 3·2 ml of 0·05 N sodium hydroxide and 5 ml of ethanol (95%). When it has dissolved, dilute the solution with 50 ml of water and 200 ml of 95% alcohol.

Method Ignite a suitable amount of sample (e.g., 10 g) in a silica dish at 450–500 °C. Add to the ash 5 ml of concentrated hydrochloric acid, heat to dissolve the ash and wash the extract into a small pear-shaped separator with water to give a total volume of 20 ml. Add 5 ml of diethylammonium diethyldithiocarbamate solution (0·8 % in chloroform) and shake for 40 s. Separate and wash with 0·5 ml of chloroform without shaking. Repeat with 2-ml portions of carbamate solution (and chloroform washes) until no more of the yellow complex of copper is formed, and finally wash with 2 ml of chloroform. Transfer the aqueous layer to a 100-ml beaker, add 5 ml of 3 N ammonium acetate solution and nearly neutralise with ammonia of sp. gr. 0·880. Add 0·2 ml of bromothymol blue and more ammonia dropwise until the solution is neutral (greenish blue) when cold. Transfer the solution to a small pear-shaped separator and wash it by shaking twice with 2 ml of chloroform. Discard the chloroform. Add 1 ml of dithizone extraction solution from a 10-ml burette and shake the mixture hard for 30 s. If it is red or pink, run off the lower layer and continue to add more of the dithizone solution until no pink colour is formed and the chloroform layer is violet or blue. When the dithizone solution is added a few drops at a time near the end-point, more pure chloroform should be added also. When titrating the blank, the initial addition of dithizone should be less than 1 ml. Lead is probably measured as well as the zinc, but in most instances the amount present is negligible.

Standardisation of dithizone extraction solution

Titrate the solution blank as above and when the violet–blue stage is reached wash out all the dithizone with chloroform. Add 1 ml of standard zinc solution and titrate the solution as above with the dithizone.

OTHER METHODS FOR DETERMINING ZINC

For reference purposes, the SAC[29] recommend that zinc is determined titrimetrically with dithizone, but prescribe a spectrophotometric finish for amounts of zinc up to 10 ppm. Previously, Sylvester and Hughes[30] described a useful method in which various metals are first extracted as dithizonates in chloroform at pH 4·5. The zinc is removed from the organic layer with 0·5 N HCl and titrated with ferrocyanide or thiosulphate according to the amount present.

TRACE ELEMENTS

REFERENCES

1. MONIER-WILLIAMS, G. W., *Trace Elements in Food*, 2nd Impression, Chapman & Hall, London (1950)
2. SOCIETY FOR ANALYTICAL CHEMISTRY, *Analyst, Lond.*, **85**, 643 (1960)
3. *Method for the Determination of Arsenic*, BS 4404:1968, British Standards Institution, London
4. HOFFMAN, I. and GORDON, A. D., *J. Ass. off. agric. Chem.*, **46**, 245 (1963)
5. SOCIETY FOR ANALYTICAL CHEMISTRY, *Analyst, Lond.*, **85**, 620 (1960)
6. *British Pharmacopoeia*, Pharmaceutical Press, London, 1242 (1968)
7. SOCIETY FOR ANALYTICAL CHEMISTRY, *Determination of Trace Elements, with Special Reference to Fertilisers and Feeding Stuffs*, Heffer, Cambridge (1963)
8. PATTON, J. and REEDER, W., *Analyt. Chem.*, **28**, 1026 (1956)
9. MOSS, P., *J. Sci. Fd Agric.*, **12**, 30 (1961)
10. POTTER, E. F. and LONG, M. C., *J. Ass. off. analyt. Chem.*, **49**, 812 (1966)
11. SAWYER, R., TYLER, J. F. C. and WESTON, R. E., *Analyst, Lond.*, **81**, 362 (1956)
12. SOCIETY FOR ANALYTICAL CHEMISTRY, *Analyst, Lond.*, **88**, 253 (1963)
13. CHENG, K. L. and BRAY, R. H., *Analyt. Chem.*, **25**, 655 (1953)
14. ABBOTT, D. C. and POLHILL, R. D. A., *Analyst, Lond.*, **79**, 547 (1954)
15. ANDREWS, J. S. and FELT, C., *Cereal Chem.*, **18**, 819 (1941)
16. PRINGLE, W. J. S., *Analyst, Lond.*, **71**, 490 (1946)
17. HAMENCE, J. H., *Analyst, Lond.*, **57**, 622 (1932)
18. SOCIETY FOR ANALYTICAL CHEMISTRY, *Analyst, Lond.*, **84**, 127 (1959)
19. *The Fertilisers and Feeding Stuffs Regulations 1968*, SI 1968, No. 218, H.M.S.O., London
20. DONALD, R., SCHWEHR, E. W. and WILSON, H. N., *J. Sci.Fd Agric.*, 7, 677 (1956)
21. WILSON, H. N., *Analyst, Lond.*, **76**, 65 (1951)
22. TSCHOPP, E. and TSCHOPP, E., *Helv. Chim. Acta*, **15**, 793 (1932)
23. FOGG, D. N. and WILKINSON, N. T., *Analyst, Lond.*, **83**, 406 (1958)
24. KIRK, R. S. and POCKLINGTON, W. D., *Analyst, Lond.*, **94**, 71 (1969)
25. SOCIETY FOR ANALYTICAL CHEMISTRY, *Analyst, Lond.*, **93**, 414 (1968)
26. RAVEN, T. W., *Analyst, Lond.*, **87**, 827 (1962)
27. BOARD, P. W. and ELBOURNE, R. G. P., *Fd Preserv. Q.*, **24**, No. 3 and 4, 53 (1964); **26**, No. 2–4, 47 (1966)
28. *Methods of Sampling and Testing Gelatines*, BS 757:1959, British Standards Institution, London
29. SOCIETY FOR ANALYTICAL CHEMISTRY, *Analyst, Lond.*, **92**, 324 (1967)
30. SYLVESTER, N. D. and HUGHES, E. B., *Analyst, Lond.*, **61**, 734 (1936)

5

OIL VALUES AND RANCIDITY

The examination of oils for identity, purity and freshness can involve a very extensive series of physical and chemical tests[1-4]. This may include measurements of density, refractive index and slip point, the methods for which are given in the official publications. The latter also describe qualitative tests for identification, various oil values, some rancidity tests and examination for metals, which may promote taints or other forms of deterioration. Physical constants and chemical values for numerous oils and fats are given by Williams[4].

The material selected for this Chapter covers the assessment of colour, the determination of the main oil values (those for iodine value, saponification value and unsaponifiable matter) together with some tests that are useful for assessing freshness or rancidity. The examination of oils for the presence of antioxidants is given in Chapter 3.

COLOUR AND SELECTED OIL VALUES

METHOD 5.1. ASSESSMENT OF THE COLOUR OF OILS AND FATS

The colour of oils and fats has a bearing on the appearance of the final product, in particular, and its assessment is of some importance industrially.

Procedure

During matching, oil samples should be at room temperature. Fats should be melted. If a fat is cloudy, filter it in an oven at 60 °C (not higher). The colour can be assessed visually or spectrophotometrically:

(*a*) Compare the colour against the glasses of a Lovibond Tintometer using a suitable cell. The colour of solidified fats can also be assessed in the porcelain trays and using the Tintometer in the vertical position.

(*b*) Using a 0·5–5 cm cell, measure the wavelength of maximum absorbance against carbon tetrachloride in a spectrophotometer. For subsequent samples (see Chapter 1), measure the optical density using the same conditions and instrument and compare the result with those obtained from earlier deliveries.

METHOD 5.2. DETERMINATION OF THE IODINE VALUE OF AN OIL OR FAT BY THE WIJS METHOD

Definition

The iodine value denotes the percentage by weight of halogen, calculated as iodine, absorbed under the conditions of the test.

Principle of method

Using standardised conditions, the glycerides of the unsaturated fatty acids present in the oil unite with a definite amount of halogen contained in Wijs' iodine monochloride solution. The degree of absorption is estimated by titration of the unused iodine with thiosulphate. Each oil and fat falls within certain ranges of iodine values, which therefore assist in identification; for example, solid fats have lower values than the more unsaturated liquid oils (see *Table 5.1*)[1, 4]. The procedure given below involves the use of only about half the amounts prescribed in the official publications.

Wijs' reagent Dissolve 8 g of iodine trichloride in about 200 ml of glacial acetic acid and mix with a solution containing 9 g of iodine dissolved in 300 ml of carbon tetrachloride. Dilute the mixture to 1000 ml with glacial acetic acid.

Procedure

Obtain two dry, stoppered reagent bottles, A and B, of 250–400 ml capacity. Pour some of the oil into a small beaker containing a small rod and weigh it to four decimal places. Transfer the sample by difference with the assistance of the rod into A and weigh it again. The approximate weight of sample to be taken in grams can be calculated from the following expression:

$$\frac{10}{\text{highest expected iodine value}}$$

For most oils and fats, the weight of sample to be used is between 0·12 and 0·24 g, which is equivalent to 4–6 reasonably sized drops from the end of the rod. Then add 5 ml of carbon tetrachloride from a dry measuring cylinder into both A and B, ensuring that the sample dissolves (if A contains solid fat, re-melt it by warming it before the addition). All the subsequent operations are performed on both A and B.

Mix and add exactly 10·0 ml of Wijs' solution (a) from a pipette (plugged with cotton-wool between the mark and the top). Swirl the solution round and insert the stopper, which should have been previously moistened with 10% potassium iodide solution, in the bottle. Allow it to stand in the dark for 30 min, then add 10 ml of 10% KI solution and 50 ml of distilled water. Swirl the solution and titrate it carefully with 0·1 N sodium thiosulphate solution. From time to time during the titration, insert the stopper and shake the bottle. When the aqueous layer becomes a very pale yellow colour *after shaking,* add starch solution and continue the titration. Just before the end-point is reached, insert the stopper *after the addition of each drop* and shake the bottle.

If the volumes of 0·1 N thiosulphate used are A ml (sample) and B ml (blank):

$$\text{Iodine value} = \frac{(B-A) \times 0 \cdot 01269 \times 100}{\text{wt. of oil or fat taken (g)}}$$

Compare the result with those given in *Table 5.1*.

Table 5.1. IODINE VALUES OF CERTAIN OILS AND FATS

Type of oil or fat	Oil or fat	Iodine value
Animal fat	Butter fat	26–40
Animal fat	Beef fat	35–44
Non-drying vegetable oil	Olive oil	79–88
Non-drying vegetable oil	Groundnut oil	85–105
Semi-drying vegetable oil	Cottonseed oil	103–113
Semi-drying vegetable oil	Sesame oil	103–116
Semi-drying vegetable oil	Soya oil	129–143
Drying vegetable oil	Linseed oil	175–200

Note
Typical values of numerous other oils and fats have been tabulated by Williams[4].

Note

(a) It is especially important to deliver the Wijs' solution from the same pipette and with exactly the same technique into both A and B. In both cases, allow the liquid to drain from the pipette for the same time and include the same number of separate drops that fall from the jet after the main flow of solution has stopped.

METHOD 5.3. DETERMINATION OF THE SAPONIFICATION VALUE OF AN OIL OR FAT

Definition

The saponification value denotes the weight of potassium hydroxide in milligrams required to saponify 1 g of the oil or fat.

Principle of method

The oil is saponified by heating it with excess alcoholic caustic alkali. The amount of alkali consumed is estimated by back-titration with hydrochloric acid. The saponification value is inversely proportional to the mean of the molecular weights of the fatty acids in the glycerides present in the oil or fat. As many oils give similar values (*Table 5.2*), the saponification value is less valuable than the iodine value for identifying an unknown oil. The main exceptions

Table 5.2. SAPONIFICATION VALUES OF CERTAIN OILS AND FATS

Oil or fat	Saponification value
Sesame oil	188–195
Groundnut oil	188–196
Olive oil	190–195
Cottonseed oil	190–198
Lard	192–200
Beef fat	194–200
Butter fat	222–232
Palm kernel oil	245–255
Coconut oil	245–265

Note
Typical values of numerous other oils and fats have been tabulated by Williams[4].

122

to this are the higher values given by coconut oil and palm-kernel oil (both used in margarine) and butter fat.

Apparatus

Conical flasks of 300–500 ml capacity fitted with water-cooled reflux condensers or long air condensers (at least 110 cm long).

Reagent

Alcoholic potassium hydroxide Dissolve 40 g of KOH pellets in 20 ml of water and dilute to 1 litre with 95 % v/v alcohol. Allow the solution to stand for a day, then filter it. The concentration should be about (but not less than) 0·5 N.

Procedure

Weigh out accurately about 2 g of the oil or melted fat into flask A. All the subsequent operations are performed on both A and the blank (B).

Add *exactly* 25·0 ml of the approximately 0·5 N alcoholic KOH solution, attach the reflux condenser and immerse the flask in boiling water for 60 min. Swirl the flask frequently during the heating. After the refluxing, add 0·5 ml of 1 % phenolphthalein and titrate very carefully, whilst still hot, with 0·5 N HCl (accurately standardised). Retain A if the unsaponifiable matter is also to be determined. If the volumes of 0·5 N HCl used are A ml (sample) and B ml (blank):

$$\text{Saponification value} = \frac{(B-A) \times 28\ 05}{\text{Wt. of oil or fat taken (g)}}$$

METHOD 5.4. DETERMINATION OF THE UNSAPONIFIABLE MATTER OF OILS

Definition and principle of method

The unsaponifiable matter consists of the substances in oils and fats (other than substances of low b.p., free fatty acids and mineral matter), which, after saponification and extraction with diethyl ether, remain non-volatile on drying at 80 °C. It includes hydrocarbons and higher alcohols. Some unsaponifiable matter (usually less than 2%) is naturally present in most oils and fats. Although a separate technique is described in BS 684[2] and the BP[1], the method below as applied to the liquid obtained after titrating the saponification value is usually adequate for routine purposes.

Procedure

Having determined the saponification value (*Method 5.3*), make the titrated liquid in flask A alkaline again with at least 1 ml of 0·5 N alcoholic KOH, transfer it to a separator and wash it in with water (50 ml less the volume of 0·5 N HCl used). Extract the solution, while still just warm, three times with 50 ml volumes of diethyl ether (the first portion should be used to wash out the original flask). Pour each ethereal extract into another separator containing 20 ml of water. After the third extract has been added, shake the combined ethereal extracts gently with the first 20 ml of wash water and then vigorously with two further 20-ml volumes. Wash the ethereal extract twice with 20 ml of aqueous 0·2 N KOH and twice or more with 20-ml volumes of water until the wash water is no longer alkaline to phenolphthalein. Pour the ethereal extract into a weighed flask, evaporate off the solvent, *dry at not more than 80 °C* and weigh to constant weight. Dissolve the unsaponifiable matter in neutral alcohol and titrate with 0·1 N alkali (not more than 0·1 ml should be required to neutralise the free fatty acid present).

With certain exceptions, the Mineral Hydrocarbons in Food Regulations prohibit the use of any mineral hydrocarbon in the preparation of foods. The main exceptions are dried fruit (maximum 0·5%), citrus fruit (0·1%), sugar confectionery (0·2%) and chewing gum. The more specific separation of mineral oil by column chromatography is discussed by Williams[4].

FRESHNESS AND RANCIDITY

The freshness and deterioration of oils and fats often has to be investigated in quality control. Rancidity may arise in food products due to the raw material, to faulty manufacturing practices or to storage. In general, the breakdown and rancidity of oils and fats is accelerated by heat, light and the presence of moisture or traces of metals such as copper and iron. The examination of oils for antioxidants is discussed in Chapter 3. Some practical aspects of the handling of fats in the confectionery industry have been discussed by Minifie[5]. Methods for determining the FFA, peroxide value and for carrying out the Kreis test are given below. A method for determining the TBA number is given on page 172.

METHOD 5.5. DETERMINATION OF FREE FATTY ACIDS

Principle of method

The oil is dissolved in neutral solvent and the acidity is titrated with standard alkali. The value obtained represents the extent to which the glycerides in the oil have been decomposed by lipase. The free fatty acids (FFA) are usually calculated as oleic acid (cf. BS 684: 1958).

Procedure

Mix 20 ml of alcohol (95 %) with 20 ml of diethyl ether, add 1 ml of phenolphthalein indicator (1 % in alcohol) and neutralise the mixture by adding 0·1 N NaOH from a burette. Weigh out 5 g of sample, using a semi-accurate balance, into a beaker-flask (200–400 ml capacity). Then add the solvent to the oil or melted fat, swirl and titrate with 0·1 N NaOH, shaking constantly, until a pink colour persists for 15 s. If two layers separate, repeat the titration using a smaller amount of sample (see also *Method 5.8*).

If titration $= V$ ml of 0·1 N of NaOH and $W =$ weight of sample taken (g), then FFA (as oleic acid) $= \dfrac{V \times 0\cdot0282 \times 100}{W}$.

Maximum limits of edibility vary considerably according to the oil, but a critical limit of 1 % can be taken as a general guide. Maxima, calculated as the acid value (which is conveniently calculated by doubling the FFA as oleic acid), are prescribed for many oils in the BP, e.g., olive oil 2·0, groundnut oil 0·5, almond oil 2·0.

During storage, the FFA of oils and fats usually increases steadily (*Figure 5.1*). The rate is, however, inhibited as the temperature is lowered.

METHOD 5.6. DETERMINATION OF PEROXIDE VALUE BY LEA'S RAPID METHOD

Principle of method

During the storage of oils and fats, oxygen is absorbed at the

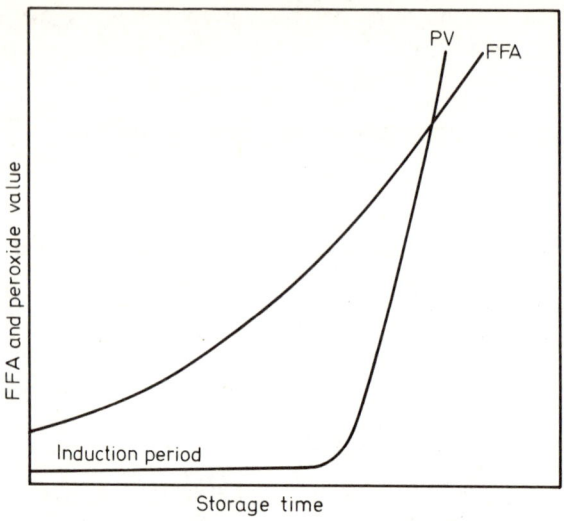

Figure 5.1. Graphs showing comparative trends of the free fatty acids and peroxide values of oils and fats during storage

unsaturated bonds, which react like those in peroxides. At a certain level, volatile products are formed that have a deleterious effect on the taste and odour, known as *oxidative rancidity*. In the usual methods, the sample is dissolved in an acetic acid–chloroform mixture and potassium iodide is added. Peroxide oxygen liberates iodine from the KI and is titrated with thiosulphate.

Procedure

Accurately weigh about 1 g of oil or melted fat into a boiling-tube, add 1 g of KI and 20 ml of solvent mixture (2:1 glacial acetic acid–chloroform) and place in *boiling* water for 60 s. Immediately, pour the hot liquid into a flask containing 20 ml of 5 % KI and wash out the tube with 15- and 10-ml volumes of water. Add starch solution and titrate with 0·002 N sodium thiosulphate (titre = T ml, which should not exceed 10 ml)

$$\text{Peroxide value (ml of 0·002 N sodium thiosulphate/g)} = \frac{T}{W}$$

where W = weight of sample taken (g).

126

Or,

Peroxide value (milliequivalents of peroxide oxygen/kg) $= \dfrac{2T}{W}$.

During the storage of most oils and fats, the peroxide value shows little increase during the early stages, known as the induction period, after which there is a marked increase (*Figure 5.1*). Therefore, although fresh oils often give no titration, a comparatively low peroxide value of, say, 3 ml of 0·002 N thiosulphate per gram is likely just to precede a marked increase, indicating oxidative rancidity (see also *Method 5.8*). Values of the order of 10–20 are usually synonymous with rancidity.

METHOD 5.7. KREIS TEST FOR INCIPIENT RANCIDITY

Principle of method

Phloroglucin reacts with oxidised fat in acidic solution to produce a red colour, the intensity of which increases during spoilage, probably owing to the presence of either malonaldehyde or epihydrin aldehyde. The method below is a sensitive qualitative test. Quantitative procedures have been described by Walters et al.[6] and Pool and Prater[7] (see also *Method 5.8*).

Procedure

Shake 10 ml of oil or melted fat in a boiling-tube with 10 ml of 0·1 % phloroglucin in diethyl ether and 10 ml concentrated HCl. The production of a pink color indicates incipient rancidity. As some fresh oils give a slight pink colour, the results have to be treated with caution.

METHOD 5.8. APPLICATION OF SEVERAL FAT SPOILAGE VALUES TO A COMMON CHLOROFORM EXTRACT[8]

Principle of method

A chloroform extract is prepared by cold maceration, so that the fat undergoes little change. Portions of the same solution are used for various spoilage values. By taking a relatively large sample, sampling errors are reduced and the results tend to be more consistent.

Procedure

(a) *Preparation of chloroform extract*—For 'pure' oils and fats, dissolve a suitable weighed amount (according to the extent of rancidity) of the oil or melted fat in chloroform and dilute the solution to a known volume. Filter it if necessary.

With fatty samples such as meat, fatty fish, butter and salad cream, place 30–150 g of sample (according to the freshness or otherwise) in a mechanical blender and add about 250 ml of chloroform. Blend the mixture for 2–3 min and filter it immediately through a large fluted paper. Then re-filter it through a paper containing a small amount of anhydrous sodium sulphate. Use portions of this second filtrate A for (b) to (f) below.

(b) *Weight of fat in the solution*—Pipette 10 ml of filtrate A into a weighed metal dish, remove the solvent, dry at 100 °C, cool in a desiccator and weigh. Use this weight for calculations in the methods below.

(c) *FFA*—Neutralise 25 ml of 95% alcohol with a few drops of 0·1 N NaOH after adding phenolphthalein. Then add this solution to 25 ml of A and titrate it with 0·1 N NaOH until the pink colour persists for 15 s. Calculate the FFA as oleic acid as a percentage of the oil.

$$1 \text{ ml of } 0·1 \text{ N NaOH} \equiv 0·0282 \text{ g of oleic acid}$$

(d) *Kreis test*—To 5 ml of filtrate (containing 1–1·5 g of fat) add 5 ml of trichloroacetic acid (30% in glacial acetic acid) and 1 ml of phloroglucinol solution (1% in glacial acetic acid) Stir the mixture by bubbling air through it for 2–3 s. Place the mixture in a water-bath at 45 °C for 15 min and add 4 ml of alcohol (95%). Measure the optical density in a 1-cm cell at 545 nm against a blank *prepared at the same time* (as for the sample except that the phloroglucin is replaced by 1 ml of glacial acetic acid). Express the result as the ratio of the optical density to the weight of oil taken.

(e) *Peroxide value—Method I*—Pipette 25 ml of filtrate into a 125-ml stoppered conical flask, add 37 ml of glacial acetic acid and 1 ml of freshly prepared saturated potassium iodide solution. Allow the solution to stand with occasional swirling for exactly 1 min, then add 30 ml of water and titrate with 0·01 N sodium thiosulphate using starch as indicator. Calculate the peroxide value as:

(a) ml of 0·002 N sodium thiosulphate/g

(b) $\dfrac{\text{titration} \times N \times 1000}{\text{wt. of sample}} = \text{mequiv./kg,}$

where N = normality of sodium thiosulphate.

(*f*) *Peroxide value—Method II* (*Sully⁹ apparatus*)—Fit a 100-ml round-bottomed flask with a ground-glass joint to a plain reflux tube about 75-cm long and of 9-mm internal diameter, the upper 15 cm of which are cooled with a water jacket (*Figure 5.2*). The flask is heated on a gauze with a semi-micro burner.

Add through the condenser into the flask 30 ml of glacial acetic acid and 5 ml of chloroform and boil the solution to the top of the tube. Whilst refluxing the solution, pour 1 g of potassium iodide (dissolved in 1–3 ml of water) down the tube. Add 25 ml of sample filtrate A and turn off the cooling water so that all the sample passes into the flask. Turn the condenser water on again. Boil the mixture for 3–5 min, cool, dilute with 50 ml of water and titrate with 0·01 N thiosulphate using starch as indicator. Calculate the peroxide value as in (*e*).

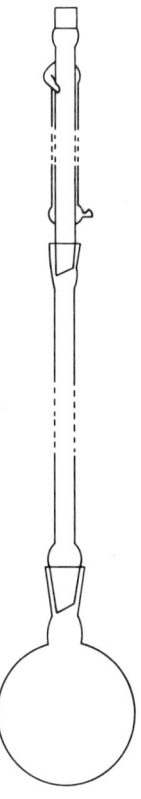

Figure 5.2. Sully reflux condenser for the estimation of peroxide values

REFERENCES

1. *British Pharmacopoeia*, Pharmaceutical Press, London, 1968
2. *Methods of Analysis of Oils and Fats*, BS 684:1958, British Standards Institution, London
3. IUPAC, *Standard Methods of the Oils and Fats Division*, 5th edn, Butterworths, London (1964)
4. WILLIAMS, K. A., *Oils, Fats and Fatty Foods*, 4th edn, Churchill, London (1966)
5. MINIFIE, B. W., *Confect. Prod.*, **36** (II), 682 (1970)
6. WALTERS, W. P., MUERS, M. M. and ANDERSON, E. B., *J. Soc. chem. Ind., Lond.*, **57**, 53T (1938)
7. POOL, M. F. and PRATER, A. N., *Oil & Soap*, **22**, 215 (1945)
8. PEARSON, D. *J. Assoc. Publ. Analysts*, **3**, 76 (1965)
9. SULLY, B. D., *Analyst, Lond.*, **79**, 86 (1954)

6

DAIRY PRODUCTS

Milk — Cream — Evaporated and Condensed Milk — Milk Powder — Butter — Margarine — Cheese — Ice Cream

MILK

Milk is examined by various physical, chemical and microbiological tests[1]. Milk arriving at the creamery or factory in bulk or churns would normally be examined for amount, smell, temperature, density, fat content, acidity and by the resazurin test. Samples with the lowest solids-not-fat (SNF) levels are also examined for added water by carrying out the freezing-point test[2]. After pasteurisation by heating at about 145–162 °F, the milk is more particularly examined by the official phosphatase test, for adequate heat-treatment and the official methylene blue test as an indication of keeping quality[3]. Also, non-homogenised milks are tested for cream line[1]. After sterilisation at not less than 212 °F, the milk is examined by the official turbidity test to ensure that the milk has been heated sufficiently to denature the albumin[3]. After ultra heat treatment (UHT) at a temperature of at least 270 °F for not less than 1 s, the milk has to satisfy a colony count test[3]. In view of the presumptive legal standards[4] (p. 137), all of these products should be regularly examined also for fat and SNF. Milk for the manufacture of many products has to be standardised at specified levels and requires checking for fat and SNF contents after blending. Also, in view of the provisions contained in an amendment to the Food and Drugs (Milk) Act 1970, UHT milk (particularly if the temperature is raised by steam injection) must have the same composition after processing as it has before the heat treatment.

Only certain selected tests are given below. Other methods mentioned are given in references 1, 2, 3, 5 and 11. The main microbiological methods are described in detail in a British Standard[6]. Bacteriological standards are officially prescribed in Scotland[7].

METHOD 6.1. SAMPLING OF MILK

Various plungers, agitators and dippers are available for the preparation of representative samples from tankers, churns, etc.[1,8]. The methods for handling and storing milk samples in relation to designated milks are prescribed in the regulations[3]. For general analyses, the sample bottles should be nearly, but not completely, filled and mixed *prior to every test* by steady continuous inversion or using a small perforated disc. Both methods keep the froth down to a minimum. *Vigorous shaking of the bottle encourages frothing and should be avoided.*

METHOD 6.2. ESTIMATION OF DENSITY OF MILK

If the milk is freshly drawn, it should be pre-warmed to 40 °C and then cooled to 20 °C. This is necessary as the density changes during the first few hours (Recknagel's phenomenon). Aged milks do not require such preparation.

The density of milk is measured by a specially calibrated hydrometer (lactometer) covering the appropriate range of specific gravities from 1·025 to 1·035, which for simplicity are marked as 25–35°. In the method, pour the mixed milk gently into the lactometer jar and allow the lactometer to slide gently into the milk. Top up the jar if necessary and take the reading on the lactometer to the nearest 0·1 after it reaches equilibrium and also the temperature. Correct the lactometer reading to 20 °C (see *Method 6.5* and *Figure 6.1*).

METHOD 6.3. ROUTINE ESTIMATION OF FAT IN MILK BY THE VOLUMETRIC GERBER PROCESS

Principle of method

Milk is added to sulphuric acid (90%) in a specially graduated tube so that the casein is dissolved. After centrifuging, the separated fat is measured directly on the scale. The addition of amyl alcohol facilitates the separation of fat. Concentrated sulphuric acid causes charring of the organic matter whereas dilute acid precipitates but does not dissolve the casein. Acid of 90% concentration therefore represents the suitable compromise concentration. Special baths, racks, pipettes, etc., are available for the handling of large numbers of samples.

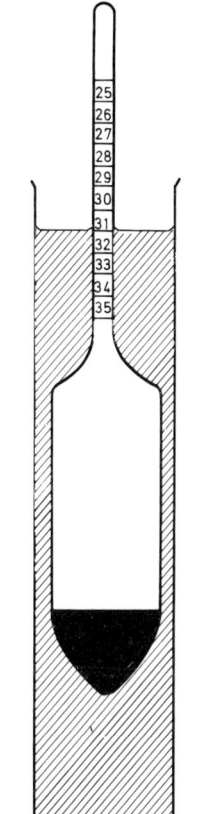

*Figure 6.1. Density lactometer
(see Method 6.5)*

Apparatus[9]

Centrifuge Specially designed to take Gerber butyrometers.
Milk butyrometer tubes Graduated up to 8% with suitable lock stoppers that are inserted with the aid of a key.
Pipettes Special pipettes marked at 10·94 ml for milk, 1 ml for amyl alcohol and 10 ml for sulphuric acid.

Reagents[10]

Gerber sulphuric acid Containing 0·895–0·910 g of H_2SO_4 per millilitre with a density of $1·815 \pm 0·003$ g/ml at 20 °C.
Amyl alcohol With a density of $0·811 \pm 0·002$ g/ml at 20 °C.

Procedure

To a milk butyrometer, held in a tall rack, carefully add the following by pipette or automatic measure:

(*a*) 10 ml of Gerber sulphuric acid;

(*b*) 10·94 ml of mixed milk;

(*c*) 1 ml of amyl alcohol.

Insert the lock stopper with the assistance of the special key and invert and shake the tube, whilst holding it in a cloth, until no white particles are seen. Check that the level of the liquid is satisfactory, i.e., the column of fat will come within the graduations after centrifuging (additions of water should be avoided if possible). Then place the tube (stopper downwards) in water at 65 ± 2 °C until the other butyrometers are ready to be transferred to the centrifuge. Centrifuge for 4–5 min at 1100 rpm and then carefully return to the 65 °C bath for 3–10 min, keeping the tube stopper downwards. Remove the butyrometer from the bath and by means of the key bring the lower end of the column of fat on to a main (whole number) graduation mark. Then read off the percentage of fat directly from the length of the column of fat to the nearest 0·05, taking the reading from the bottom of the top meniscus to the lower flat whole number position (*Figure 6.2*). With *homogenised milk* and *sterilised milk* (or, if there is any doubt, with non-homogenised milk) return the tube to the 65 °C bath, re-centrifuge and read again. Wash the tube out with hot water as soon as possible after the determination and place it in the tall rack to drain.

As a check for reference purposes, the fat should be determined by the gravimetric Rose–Gottlieb process (*Method 6.4*).

METHOD 6.4. ESTIMATION OF FAT IN MILK BY THE ROSE–GOTTLIEB METHOD[11]

Principle of method

Alcohol is added to the milk to precipitate the casein, which is dissolved in ammonia. The fat is first extracted with diethyl ether, which is a very efficient solvent. Light petroleum is also added as it reduces the solubility of non-fatty materials such as lactose, which may be included in the diethyl ether extract. The method is preferable to methods in which acid is used as no heat is applied during the extraction. It is also more suitable for materials that contain much sugar, such as sweetened condensed milk.

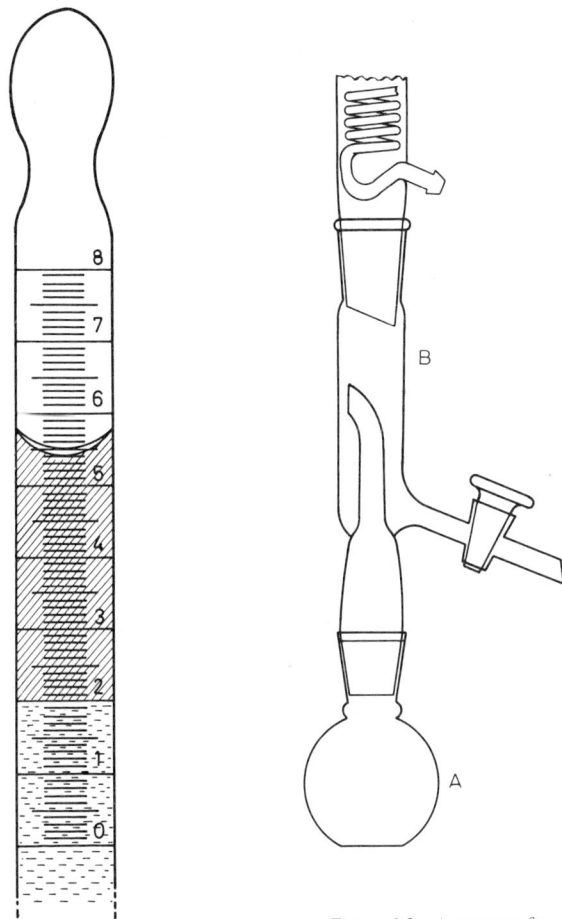

Figure 6.2. Reading of Gerber milk butyrometer (see Method 6.5)

Figure 6.3. Apparatus for removing solvent subsequent to extraction of fat

Apparatus

Mojonnier-type fat extraction apparatus (*Figure 3.3*—p. 87). Apparatus B for removing solvent with flask A (*Figure 6.3*).

Procedure

To save time, dry the flask A (required later) at an early stage in a

135

100 °C oven for 30 min, cool it in a desiccator and weigh it.

Weigh out accurately about 10 g of milk into a Mojonnier tube (*Figure 3.3*), add 1 ml of ammonia of sp. gr. 0·880 and mix well. Add 10 ml of 95% alcohol and mix again by allowing the liquid to flow backwards and forwards. Then add 25 ml of diethyl ether, close the tube tightly with a cork (previously wetted with water) and shake it vigorously for 1 min. Then add 25 ml of light petroleum (boiling range 40–60 °C) and shake the mixture vigorously for 30 s.

Allow the tube to stand for at least 30 min to effect complete separation of the layers (*a*). Then decant as much as possible of the upper layer into flask A. Repeat the extractions at least twice using 15 ml of diethyl ether and 15 ml of light petroleum and similarly pour the extracts into A. Finally, fit flask A with B and attach the water condenser. Heat A in hot water (at approximately 80 °C) and tap off the mixed solvent from B. Remove flask A, wipe it carefully, dry it for 45–60 min in an oven at 100 °C, cool it in a desiccator and weigh it (*b*). Calculate the percentage of fat in the original sample (cf., BS 1741[11]).

Note

(*a*) If the interface between the two layers is below the lower end of the narrow part of the tube, raise it to this level by adding carefully a small volume of water down the side of the tube. Emulsions can often be broken by adding a small volume of alcohol.

(*b*) If non-fatty matter derived from the aqueous layer is present, the warmed fat should be completely washed out from the flask with several small volumes of light petroleum. The amount of fat is then obtained by difference by weighing the flask after re-drying it in the oven.

METHOD 6.5. CALCULATION OF TOTAL SOLIDS AND SOLIDS-NOT-FAT FROM THE MODIFIED RICHMOND EQUATION

Milk consists essentially of a mixture of water (sp. gr. 1·0), fat (sp. gr. 0·93) and SNF (sp. gr. 1·61). The proportions of each vary, but most milk has a specific gravity from about 1·029 to 1·033 (corresponding to 29–33° as a lactometer reading). Richmond found that it was possible to calculate the total solids from the lactometer and fat values. The modified current relationship (BS 734:1959[12]) is:

$$\text{Total solids} \, (\%) = T = 0 \cdot 25 \, D + 1 \cdot 22 \, F + 0 \cdot 72$$

where D = density lactometer reading at 20 °C and F = percentage of fat. The temperature correction for D is 0·24 units for each 1 °C change in temperature.

(a) Estimation of solids-not-fat using a special slide-rule

Although tables are given in BS 734:1959[12] based on the modified Richmond equation, it is more convenient to use a special slide-rule for the purpose. Its use can best be explained by taking a typical example. Suppose, therefore, that a sample gives the following readings (*Figures 6.1* and *6.2*):

BS lactometer reading at 18 °C = 31·0°

Fat (Gerber method) = 3·45%

Note that for routine purposes the fat, TS and SNF are normally estimated to the nearest 0·05%.

(b) Temperature correction for density

Hold the rule upright so that the arrow on the central moving slide (CS) points to the left (*Figure 6.4*). Move the slide so that the lactometer reading 31·0 CS coincides with 20 °C (not 18 °C) on the temperature scale on the lower left-hand side of the fixed part of the rule. Then read off the lactometer reading of 30·5 corresponding to 18 °C on the temperature scale.

(c) Estimation of solids-not-fat

Hold the rule horizontally with the arrow on CS pointing upwards. Noting that the fat scale (top right) goes from right to left, make the arrow on CS coincide with 3·45% on the fixed fat scale (*Figure 6.5*). Then read off the total solids (12·55%) on the lower fixed scale corresponding to the density reading at 20 °C on CS of 30·5.

		Presumptive minima
Total solids (%)	= 12·55	—
Fat (%)	= 3·45	3·0
Solids-not-fat (%)	= 9·10	8·5

With doubtful samples, the total solids should be checked by accurately weighing out about 5 g of mixed milk (add 5 ml from a

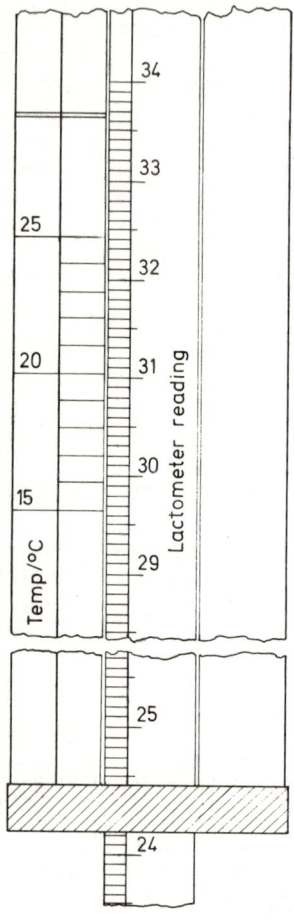

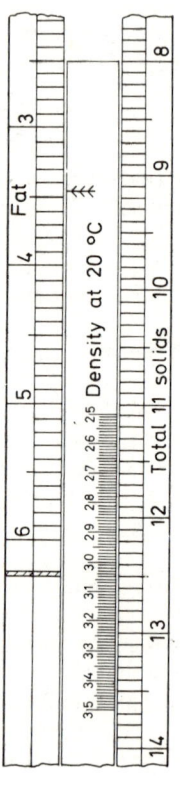

Figure 6.4. Correction of density reading of milk for temperature using slide-rule (see Method 6.5)

Figure 6.5. Calculation of total solids of milk using slide-rule (see Method 6.5)

pipette) into a metal dish and drying first on a boiling water bath. After about 30 min, wipe the outside of the dish and place it in an oven at 100 ± 1 °C for $2\frac{1}{2}$h. Then cool it in a desiccator and weigh it. Continue the drying at 100 °C for 1-h periods until constant weight is achieved. The reference method for fat is the Rose–Gottlieb method (*Method 6.4*).

METHOD 6.6. ASSESSMENT OF ADDED WATER IN MILK FROM THE DETERMINATION OF THE FREEZING POINT BY THE HORTVET METHOD

The f.p. of milk as it comes from the cow usually falls between narrow limits, viz., from -0.530 °C to -0.550 °C. If water is added the f.p. gets nearer to that of water (0.0 °C). The proportion of water added can be readily assessed by simple proportion from the f.p. of the sample. The test is normally applied to milks with SNF below 8.5 % or to ones from sources that are under suspicion. The standardised technique outlined below is the BS modification of that originally devised by Hortvet. Full details are given in BS 3095[2].

Apparatus

The milk is frozen in a glass tube that slides down into a metal tube, which is surrounded with diethyl ether. The ether can be re-used and should be stored cold in the refrigerator. The ether is contained in a Dewar flask and cooled by blowing air (dried by passage through tubes containing concentrated sulphuric acid) through it throughout a series of determinations. For reasons of safety, it is important that a drain tube is used to convey the ether vapour out of the window and not into the laboratory. The temperature of the ether is controlled by a thermometer reading from -5 °C to $+5$ °C. The long Hortvet thermometer that is held in the tube of milk for determining the f.p. has a short range, from -2 °C to $+1$ °C, and must be kept upright at all times. The stopper holding the Hortvet thermometer also includes two further holes:

(*a*) containing a metal tube through which passes the stirrer;

(*b*) a hole through which is inserted a metal tube containing ice for seeding the sample just prior to taking the final reading. It is advisable to retain the end of this freezing starter tube in ice until required.

Semi-automatic apparatus available incorporates a motor that operates (*a*) the pump for blowing air through the ether and (*b*) the stirrer.

The Hortvet thermometer is standardised against water and sucrose solutions, which should give the following values:

Concentration of sucrose at 20 °C, % w/v	f.p. depression/ °C	Concentration of sucrose at 20 °C, % w/v	f.p. depression/ °C
7·0	0·422	8·5	0·520
7·5	0·454	8·75	0·537
8·0	0·487		

Procedure

Pour 500 ml of cooled diethyl ether into the Dewar flask via the T-piece using a thistle funnel. Check the level by means of a glass tube, which should be marked to assist with topping up between determinations. Close the flask with a cork and pass dried air through the ether so that the control thermometer reads $-3\cdot0$ °C. Re-check the ether level and restore the volume to 500 ml. Pour a small volume of alcohol into the metal tube to give conduction with the glass tube. Into the latter pour 35–40 ml of mixed cooled milk (the acidity should preferably not exceed $0\cdot18\%$ w/v as lactic acid) or boiled and cooled water, attach the Hortvet thermometer and stirrer and insert in the metal tube. Re-pass the air through the ether so that the control thermometer reads $-3\cdot0$ °C ($-2\cdot5$ °C with water). Stir the sample at a steady up-and-down motion of 45 strokes per minute until the temperature of the milk on the Hortvet thermometer shows a supercooling of $1\cdot0$ to $1\cdot2$ °C, i.e., f.p. of $-1\cdot5$ to $-1\cdot7$ °C ($-1\cdot0$ to $-1\cdot2$ °C with water). Then insert the freezing starter tube, containing a piece of ice, through the hole in the stopper to seed the sample. Keep stirring the solution until the Hortvet thermometer shows a rise in temperature and then remove the starter tube. Then stir three times as the mercury becomes stationary, tap the thermometer seven times and estimate the reading to $0\cdot001$ °C. About $\frac{1}{2}$ min after taking the reading, repeat the stirring, tapping and reading operations. Repeat again after a further $\frac{1}{2}$ min and record the highest of the three readings.

Calculation of proportion of added water

If the actual f.p. reading of water on the thermometer is F_w °C and the actual f.p. reading of the milk sample on the thermometer is F_s °C,

140

then the positive depression of f.p. below zero of the milk sample, Δ, is given by:

$$\Delta = -(F_s - F_w)$$

$$\text{Added water } (\% \text{ w/w}) = \frac{\Delta_1 - \Delta}{\Delta_1} \times (100 - T)$$

where T = total solids of the milk ($\% \text{ w/w}$) and Δ_1 = depression of f.p. of genuine milk.

Occasionally, the true value for Δ_1 (that for the milk from which the sample was watered) is known. Normally, however, the added water is calculated on the assumption that Δ_1 is 0·540, i.e., a suggested mean f.p. depression value for unadulterated milk. For legal purposes, however, a minimum value for Δ_1 of 0·530 is permissible. Further points of legal significance have been reviewed by the author[13].

Example of calculation

$$F_w = +0\cdot010 \,^{\circ}\text{C}$$
$$F_s = -0\cdot490 \,^{\circ}\text{C}$$
$$T = 11\cdot0\%$$

Therefore,

$$\Delta = -(F_s - F_w) = -[-0\cdot490 - (+0\cdot010)] = 0\cdot500$$

$$\text{Added water } (\%) = \frac{\Delta_1 - \Delta}{\Delta_1} \times (100 - T)$$

$$= \frac{0\cdot540 - 0\cdot500}{0\cdot540} \times 89 = 6\cdot6\%$$

(or $$= \frac{0\cdot530 - 0\cdot500}{0\cdot530} \times 89 = 5\cdot0\%)$$

METHOD 6.7. ESTIMATION OF TOTAL ACIDITY OF MILK

Pipette 10 ml of mixed milk into a porcelain basin and add 1 ml of 0·5% phenolphthalein. Titrate the mixture with 0·1 N sodium hydroxide solution and calculate the acidity as lactic acid ($\% \text{ w/v}$) (cf. BS 1741[11]).

1 ml of 0·1 N NaOH $\equiv$ 0·0090 g of lactic acid

Fresh milk normally gives a titre of 1·4–1·8 ml ($\equiv 0·13$–0·16% of lactic acid). Milk arriving at the creamery is considered to be unsatisfactory if the acidity exceeds 0·18%. Milks with slightly higher acidities are suitable, however, for the manufacture of some cheese. The pH of fresh milk is about 6·6 and falls to 4·3 when it goes sour. A low acidity in a raw milk is not necessarily synonymous with a low bacterial count.

METHOD 6.8. RESAZURIN TEST FOR MILK

Principle of method

The resazurin test is a chemical test for assessing hygienic quality. It measures the degree to which actively proliferating bacteria are able to reduce the redox indicator resazurin:

$$\text{resazurin} \rightarrow \text{resorufin} \rightleftharpoons \text{dihydroresorufin}$$

blue   pink   colourless

The degree of reduction is measured on a Lovibond Comparator disc. The method is more rapid and therefore more suitable as a platform test than the officially prescribed methylene blue test.

The usual aseptic precautions must be taken during the test, including preparation of the reagent[6].

Reagent

Resazurin Dissolve one standard resazurin tablet in 50 ml of cold sterile water in a stoppered flask. Use the solution within 4 h of its preparation.

Procedure

Aerate the sample by shaking and flame the rim of the bottle. After removing its cotton-wool plug, flame a sterilised tube marked at 10 ml and pour the mixed milk sample into it up to the mark. Then add 1·0 ml of resazurin solution from a sterile pipette, close the tube with a sterile rubber stopper (previously heated in boiling water) and slowly invert the tube twice. Place it in a covered water-bath at 37–38 °C for 1 h (invert the tube after 30 min) and match the colour against the resazurin disc of the Lovibond Comparator.

The interpretation of the results are as follows:

Disc number	Condition of milk
4–6	Satisfactory
1–3½	Doubtful
0 or ½	Not satisfactory

Using the 1-h incubation method, the disc readings correspond approximately to the number of hours required to reduce methylene blue, which is used in the statutory test[3, 14-16].

METHOD 6.9. PHOSPHATASE TEST FOR ASSESSING THE ADEQUACY OF HEAT TREATMENT OF PASTEURISED MILK[3]

Principle of method

According to the Special Designation Regulations[3], milk is adequately pasteurised if it is submitted to one of the following conditions of processing:

(a) retained at 145–150 °F for at least 30 min, or

(b) retained at not less than 161 °F for at least 15 s.

The milk is then required to be cooled immediately to not more than 50 °F.

The statutory test for adequacy of heat processing is based on the finding that phosphatase enzymes are destroyed during the heat treatment. Also, this enzyme is more difficult to destroy than tubercle bacilli. In the test, the milk is incubated with disodium p-nitrophenyl phosphate under alkaline conditions. The phosphatase present in raw milk decomposes the substrate, liberating free p-nitrophenol, which is yellow. The degree of inactivation of phosphatase is assessed by comparing the yellow colours produced with those on a comparator disc.

Special apparatus

Water-bath Maintained at 37.5 ± 0.5 °C.
Pipettes Straight-sided to deliver 1.0 ml.
Test-tubes With rubber stoppers, which should be heated in boiling water prior to use.
Lovibond 'all purposes' comparator With APTW or APTW7 disc and 25-mm cells.

Reagents

Buffer solution Dissolve 3·5 g of anhydrous $Na_2CO_3 + 1·5$ g of $NaHCO_3$ in water and dilute the solution to 1000 ml.
Substrate Disodium *p*-nitrophenyl phosphate, stored in a refrigerator.
Buffer–substrate solution Dilute 0·15 g of substrate to 100 ml with buffer solution and store the solution in a refrigerator and away from light. It should give a reading of less than 10 on the comparator disc using water for comparison.

Care of apparatus

New glassware should be cleaned by soaking it in chromic acid mixture. Test-tubes should be soaked in soda, and if necessary in HCl also. After cleaning, glassware should be rinsed with distilled water and dried. Avoid introducing saliva into the pipettes.

Procedure

Pipette 5 ml of buffer–substrate solution into a test-tube, stopper it and place it in the 37 °C bath. When the liquid reaches this temperature, add 1·0 ml of the pasteurised milk, re-stopper the tube and mix the contents by shaking. Then incubate the mixture for exactly 2 h at 37 °C together with a blank containing boiled milk (prepared from the test sample). After the incubation, mix the contents of the sample tube and place it in the right-hand side of the comparator; place the blank on the left-hand side. Match the sample colour on the disc. A reading of 10 μg or less of *p*-nitrophenol per millilitre of milk indicates that pasteurisation is satisfactory. A higher reading indicates inadequate heat treatment for pasteurisation or the presence of or contamination with raw milk.

The statutorily prescribed test for UHT milk is based on the coliform count[3].

METHOD 6.10. STATUTORY TURBIDITY TEST FOR STERILISED MILK[3]

Principle of method

The milk is shaken with ammonium sulphate, filtered and the filtrate is heated in boiling water. Any non-denatured albumin present as a result of insufficient heat treatment of the milk separates with the casein. No turbidity is produced in the filtrate if the milk has been heated for the necessary period at not less than 212 °F as prescribed in the Special Designation Regulations[3].

Procedure

Weigh out 4 ± 0.1 g of ammonium sulphate into a 50-ml conical flask, add 20 ± 0.5 ml of milk sample and shake the mixture for 1 min to dissolve the ammonium sulphate. Allow the solution to stand for at least 5 min, then filter it through a 12·5-cm Whatman No. 12 folded filter-paper into a test-tube. When at least 5 ml of clear filtrate have been collected, place the tube in a beaker of boiling water for 5 min. Then cool it in cold water and examine the solution for turbidity. A sterilised milk that has been satisfactorily heat treated gives no turbidity. UHT milk gives a faint turbidity and raw and pasteurised milk give a white precipitate.

METHOD 6.11. A TEST FOR REDUCING CAPACITY TO ASSESS THE DEGREE OF HEAT TREATMENT OF STERILISED MILK[17]

Principle of method

A negative result obtained in the official test (*Method 6.10*) merely indicates that the milk has been heated at or beyond a temperature which may be much below the boiling point of milk (and also much below the temperature and time range of commercial practice in milk sterilisation). Fellows[17] therefore recommended the use of a colorimetric test involving the use of ferricyanide, which assesses the extent to which the milk has been heated and is therefore of more value than the statutory test in industrial control. The test is based on the findings of Chapman and McFarlane[18] that (*a*) fresh milk has a considerable capacity to reduce ferricyanide and (*b*) heat treatment increases the reducing capacity, probably owing to the formation of reducing substances derived from lactose and protein[19]. As pH has been shown to have a significant effect on the reduction potential, it is important that the buffer solution is prepared with care.

Buffer solution

Add 35·6 ml of 0·2 N sodium hydroxide to 100 ml of 0·2 M potassium dihydrogen phosphate (27·232 g/l) and dilute the solution to 400 ml.

Procedure

Mix 2·5 ml of the milk sample with 2·5 ml of water in a test-tube and add 5 ml of buffer solution and 5 ml of *freshly prepared* 1 % potassium ferricyanide solution. Mix the contents and place the tube in a water-bath at 50 °C for 20 min and immediately cool it in ice-water to below 25 °C. Add 5 ml of 10 % trichloroacetic acid, mix and filter using a Whatman No. 40 filter-paper. Dilute 5 ml of filtrate with 5 ml of water and add 1 ml of *freshly prepared* 0·1 % ferric chloride solution. Mix, allow the solution to stand for 10 min and read the optical density at 660 nm. For setting the instrument, use a control tube prepared in the same way as the sample but using 5 ml of water instead of the diluted milk.

Results

Fellows applied the method to various milks that had been submitted to various degrees of heat treatment and reported values of the following order for the 'reducing capacity'. The readings were obtained with a Hilger Biochem absorptiometer using an orange filter.

Sample	Reducing capacity (log reciprocal readings)
Raw milk	0·20
Milk pre-heated to 165 °F (73·9 °C)	0·18
Homogenised milk at 168 °F (75 °C)	0·18
Sterilised milk	0·39

Ageing of samples for 24 h increased the reducing capacity of the raw milk and preheated milk, but had little effect on the sterilised milk.

CREAM

Milk can be mechanically separated to produce cream with a widely varying milk fat content[20]. The minimum fat contents prescribed in statutory regulations[21] vary from 12 to 55 %, e.g., single cream 18 %, double cream 48 %.

Cream is examined for taste, smell, water (dry 1–2 g at 100 °C), fat (see below), acidity (titrate 10 g of sample + 10 ml of water as for milk (*Method 6.7*) and calculate as per cent. w/w lactic acid), the resazurin and methylene blue tests and for microbiological examination[6, 9-11]. Keeping qualities can be assessed by titrating the acidity after incubating at 18–20 °C. The sample must be stirred thoroughly but not vigorously before analysis.

METHOD 6.12. ESTIMATION OF FAT IN CREAM BY THE GERBER METHOD USING A MILK BUTYROMETER

Apparatus and Reagents—See *Method 6.3*[9, 10].

Procedure

Weigh out 1·1 g of cream rapidly into a small porcelain dish. Pipette 10 ml of Gerber sulphuric acid into a milk butyrometer. Then, with the assistance of about 6 ml of hot water delivered from a washbottle, transfer the sample into the butyrometer down a rod via a thin-stemmed funnel. Add 1 ml of amyl alcohol and sufficient water to bring the total volume up to the usual level, stopper and complete as for milk (*Method 6.3*), centrifuging twice.

$$\% \text{ Fat} = (10 \times \text{milk butyrometer reading}) - 1$$

A butyrometer with graduations from 0 to 70% of fat, specially designed for use with 5 g of cream, is also available. In addition, the sample can be weighed into a funnel with stopper (*Figure 6.6*). On

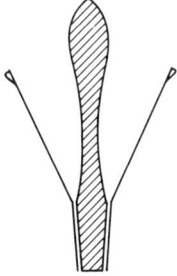

Figure 6.6. Funnel with stopper for weighing out samples such as cream and cheese into butyrometers

partially removing the stopper, the sample can be washed into the butyrometer with the requisite amount of water.

When greater accuracy is required, the Rose–Gottlieb method (*Method 6.4*) is to be preferred.

EVAPORATED AND CONDENSED MILK

Milk that has been standardised, concentrated, homogenised and sterilised in cans is referred to in the trade as evaporated milk. Viscosity is an important factor controlling foaming and foam stability. Problems associated with evaporated milk are the formation of a cooked flavour and gelation on storage. Sweetened condensed milk relies on the presence of added sucrose for its keeping qualities (rather like a preserve). A sugar ratio, viz.,

$$\frac{100 \times \% \text{ sucrose}}{\% \text{ sucrose} + \% \text{ water}}$$

of at least 62·5 should protect a condensed milk from microbiological defects provided that the raw milk used is of good bacterial quality and is adequately pre-warmed. Rapid crystallisation of small crystals is achieved by seeding with powdered lactose. The crystals in the product should be below 10 μm in size. If the size exceeds 15 μm, the crystals produce 'sandiness,' which can be detected on the tongue. The compositional quality of all types (full-cream, half-cream, skimmed; sweetened or unsweetened) are covered by regulations[22], which prescribe, for example, minima of 9% of fat and 31% of total milk solids for all full-cream condensed milk. Labels must state the equivalent pints present in the container relative to a standard milk.

Typical tests on the final product are those for net weight, viscosity (by comparison of time of flow down a funnel), colour, total solids, fat, added sugar and acidity. Reference methods are described in a British Standard[23] for total solids, sucrose (polarimetrically) and fat. For the total solids, 1·5 g of sweetened (or 3 g of unsweetened) condensed milk is weighed into a dried metal dish plus rod, containing sand. Then 3–5 ml of water is added and mixed with the sample plus sand and the dish plus contents are dried for 20 min on a boiling water bath and then to constant weight in an oven at 100 °C. In industrial control, the sugar can be readily controlled within the desired limits by using a refractometer. The fat content can be more rapidly assessed by the Gerber process (see below). The equivalent pints can be calculated from relevant equations[13].

METHOD 6.13. ESTIMATION OF FAT IN EVAPORATED AND CONDENSED MILK BY THE GERBER METHOD

Apparatus and Reagents—See Method 6.3[9,10].

Procedure

Weigh out into a 100-ml beaker 20–25 g of sample (±0.01 g) and add 20 ml of water. Stir the mixture with a clean glass rod and pour it down the rod through a funnel into a 100-ml volumetric flask. Similarly, wash out the beaker with several small volumes of water and make up the volume to 100 ml. Mix gently by frequent inversion of the flask and treat 10·94 ml of the solution by *Method 6.3*, centrifuging at least twice.

$$\text{Fat (\%)} = \frac{\text{Milk butyrometer reading}}{\text{Weight taken}} \times 100$$

The method described for ice cream (*Method 6.23*) using Macdonald's reagents can also be used for condensed milk.

The Rose–Gottlieb reference method involves adding 8 ml of warm water to 2–2·5 g of condensed milk, then mixing successively with 8 ml of warm water, 1 ml of ammonia of sp. gr. 0·880 and 10 ml of alcohol and extracting the fat as described for milk (*Method 6.4*).

MILK POWDER, DRIED MILK

Milk powder is usually prepared by spray-drying or roller-drying of standardised pre-concentrated milk[24]. Statutory regulations[25] prescribe compositional standards for six types varying in milk fat content from a minimum of 26% for dried full-cream milk to a maximum of 1·5% for dried skim milk. In general, compared with roller-drying, spray-drying gives a product with higher solubility, better appearance and flavour, but often has inferior keeping qualities. Overheating during roller-drying is liable to produce burnt particles. With foam drying and freeze drying the changes in the milk constituents is minimal.

Apart from standards for fat contents, legal requirements also include a maximum of 5% of moisture and provision for labels, including the statement of the equivalent pints relative to various standard milks.

Reference methods of analysis are described in a British Standard[26] for moisture (dry at 102–103 °C), fat, total N, ash, alkalinity of ash, titratable acidity and bulk density. Industrially, acidity would normally be checked on the raw material, but the product is examined for colour, flavour and solubility in addition to moisture and fat. Deteriorated samples should be examined for fat spoilage values, e.g., peroxide value (*Method 5.8*). For moisture, rapid HTST ovens are not particularly accurate for dried milk and this has led to the production of special equipment, e.g., the use of a humidity sensor based on semiconductors[27]. In the factory, the moisture content is controlled at 2·5–3·5%.

METHOD 6.14. ESTIMATION OF FAT IN MILK POWDER BY THE GERBER METHOD

Apparatus and Reagents—See *Method 6.3*[9, 10] (p. 132).

Procedure

Pipette 10 ml of Gerber sulphuric acid into a milk butyrometer and carefully add cold water to form a layer 6 mm deep. Weigh out 1·69 ± 0·01 g of dried milk in a small metal scoop and add it to the tube down a dry funnel with a wide but very short stem. If, even after tapping the funnel, some powder still adheres, use the subsequent additions to complete the transference. Add by pipette 1 ml of amyl alcohol and sufficient hot water (70 °C) from a wash-bottle until the butyrometer is filled to about 5 mm below the shoulder. Stopper and mix the contents of the tube by successively shaking and inverting. Complete as for milk (*Method 6.3*), centrifuging twice.

$$\text{Fat} \, (\%) = \frac{20 \times \text{milk butyrometer reading}}{3}$$

The reference method for fat involves shaking 10 ml of water with 1 g of full-cream milk powder (1·5 g with skimmed milk powder) in a Mojonnier tube (*Figure 3.3*), warming, adding 2 ml of 25% ammonia (sp. gr. 0·91), and then 10 ml of alcohol and extracting with 25 ml of diethyl ether followed by 25 ml of light petroleum as in *Method 6.4*.

METHOD 6.15. ASSESSMENT OF SOLUBILITY OF MILK POWDER (ADMI)[28]

Principle of method

In the method recommended by the American Dry Milk Institute outlined below, the solubility of the dried milk is assessed from the sediment remaining after maceration with hot water, and centrifuging. The result is then expressed as the 'Solubility Index'. Other methods used measure the proportion of dissolved material. All methods are essentially empirical and in order to obtain truly comparative results it is important to follow the same procedure on all samples[24].

Apparatus

Centrifuge The ADMI specifies the required centrifuging speed from 695 to 1075 rpm according to diameters of head from 10 to 24 in.

Centrifuge tubes Conical graduated centrifuge tubes of volume 50 ml.

Mixer The ADMI specifies that the blades of the impeller must have a pitch of 30° and a spread between blades of 11/32 in. It is intended for use with a special jar.

N.B. Any suitable combination of mixer and jar may be used, provided that it gives suitably comparative results.

Procedure

Add 10 g of skimmed or 13 g of full-cream milk powder to 100 ml of water (at 75 °F) in the mixing jar. Add 3 drops of diglycol laurate S, stir the mixture for 90 s and allow it to stand for not more than 15 min. Then mix it thoroughly for 5's with a spoon, immediately pour it into a centrifuge tube up to the 50-ml mark and centrifuge it for 5 min. Immediately siphon off the supernatant liquid to within 5 ml of the sediment surface without disturbing the sediment. Add 25 ml of water (at 75 °F) and shake the tube gently to disperse the sediment, dislodging it if necessary with a wire. Fill the tube to the 50-ml mark with water (at 75 °F), invert it several times to mix the contents thoroughly and centrifuge for 5 min. Then hold the tube in a vertical position with the upper level of the sediment at eye level and assess the volume of sediment as accurately as possible (preferably to the nearest 0·1 ml). A strong source of light is useful for distinguishing the sediment (*Table 6.1*).

Spray-dried powders have a much superior solubility to roller-dried products.

Factors that affect the reconstitutability and dispersibility of milk powder have been discussed by King[24].

METHOD 6.16. MEASUREMENT OF BULK DENSITY OF MILK POWDER[26]

Apparatus

Measuring cylinder Capacity 100 ml with a 100-ml mark approximately 15 cm above the zero mark. The cylinder is closed with a cork.

Table 6.1. LIMITS FOR SOLUBILITY INDEX (AMERICAN DRY MILK INSTITUTE, CHICAGO)

Type of dry milk	Grade	Process	Solubility index (max.)/ml
Non-fat	Extra	Spray	1·25(a)
Non-fat	Extra	Atmospheric roller	15·00
Non-fat	Standard	Spray	2·00(b)
Non-fat	Standard	Atmospheric roller	15·00
Instant non-fat	Extra	—	1·00
Dry whole	Premium, extra	Gas packed, spray	0·50
Dry whole	Extra	Bulk, spray	0·50
Dry whole	Standard	Bulk, spray	1·00
Dry whole	Extra, standard	Bulk, roller	15·00

Notes:
(a) If designated 'high heat', maximum is 2·0 ml.
(b) If designated 'high heat', maximum is 2·5 ml.

Procedure

Transfer 20 g of milk powder to the 100-ml cylinder. Stand the cylinder on a soft pad about 1·5 cm thick (e.g., a folded duster) and then clamp it 15 cm above the pad. Allow the cylinder to drop 15 cm on to the pad ten times. Level off the powder by tapping the cylinder and calculate the bulk density as grams per millilitre.

The bulk density of spray-dried milk powder is usually 0·5–0·6 g/ml. Instant powders give a lower value, of the order of 0·35 g/ml.

BUTTER

Butter is made by churning cream separated after allowing milk to stand in a warm place[29]. Excess water is removed and up to 2% of salt is worked in to form a homogeneous mass. Flavoured butter made from sour or ripened cream contains more diacetyl and has a better flavour than sweet cream butter made without starter. Regulations[30] prescribe maxima of 16% of water and 2% of milk solids other than fat and a minimum of 80% milk fat (or 78% for a very salt butter). The raw material used may be partially neutralised with alkali carbonate. Other permitted additions are the colours annatto, carotene and turmeric.

Reference methods are described in a British Standard[31] for moisture (dry at 100 °C to constant weight), curd and salt, fat (by difference, i.e., 100 − % moisture − % curd − % salt), titratable acidity, copper, iron, pH of the serum and examination of the fat for the Reichert, Polenske and Kirschner values, iodine value, acidity, saponification value and refractive index. Industrially, the main methods are those for flavour, rheological tests (e.g., penetrometer), moisture, fat (on 2·5 g in a 5-g cream Gerber or van Gulik butyrometer), salt (by direct titration—see below) and pH. The proportion of butter fat present is frequently assessed from the Reichert and Polenske values. Rancidity tests can be conveniently performed using *Method 5.8*. Sampling methods for butter and butter fat are included in the relevant British Standards[8, 31].

METHOD 6.17. DIRECT DETERMINATION OF SALT IN BUTTER

Using a semi-accurate balance, weight out 3 g of sample in a porcelain evaporating basin (approximately 7–9 cm in diameter) and add 15 ml of boiling water (or heat on a water-bath until the fat melts). Stir the mixture with a rod and, after allowing it to cool, add 0·5 ml of 5% potassium chromate. Titrate with 0·05 N silver nitrate to a *faint orange* end-point, stirring continuously.

$$1 \text{ ml of } 0·05 \text{ N AgNO}_3 \equiv 0·00292 \text{ g of NaCl}$$

METHOD 6.18. ESTIMATION OF THE PROPORTION OF BUTTER FAT BY THE SEMI-MICRO REICHERT AND POLENSKE PROCESSES

Principle of method

The *Reichert value* is the volume of 0·1 N alkali in millilitres required to neutralise the water-soluble volatile fatty acids distilled from 5 g of fat under specified conditions. From the same distillation, the *Polenske value* is the volume of 0·1 N alkali in millilitres required to neutralise the water-insoluble volatile fatty acids.

The reference method using 5 g of fat is described in BS 769[31]. The more rapid semi-micro process using 1 g of fat and titrating the distillate with 0·02 N alkali described below has been shown to give similar results and is usually more convenient for assessing the proportion of butter fat in the fat extracted or rendered from products.

In the process, the fat is saponified with sodium hydroxide. Water is added, and the resulting soap solution is acidified with sulphuric acid and distilled under standardised conditions. After filtration, the water-soluble volatile fatty acids are titrated in the filtrate (the Reichert value). The washed insoluble volatile acids are dissolved off the filter with neutral alcohol and titrated (the Polenske value).

Apparatus

Pipette Re-calibrate a 1-ml graduated pipette with a mark to indicate a volume equivalent to 1·0 g of butter fat.

Semi-micro distillation apparatus With distillation flask A (capacity

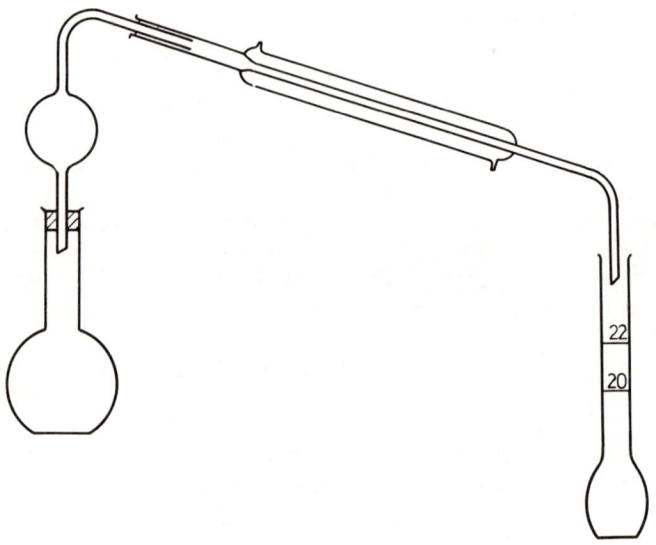

Figure 6.7. Semi-micro Reichert distillation apparatus

60 ml), condenser (approximate length 20 cm, with a water jacket 13 cm in length) and receiving flask B (with marks at 20 ml and 22 ml)—see *Figure 6.7*.

Sampling procedure

Sampling of butter and margarine Heat a portion of the sample to about 60 °C in a beaker until fat separates above the layer. Then pour the upper fat layer carefully through a dried filter-paper into a warmed flask. Liquefy the fat before taking samples for analysis.

Sampling of butter confectionery Dissolve the sample in water, warm with dilute HCl, cool and extract the fat with a diethyl ether–light petroleum mixture. Remove the solvent and dry the fat at 100 °C.

Sampling of cream soups Extract the fat directly with a diethyl ether–

light petroleum mixture and continue as above for butter confectionery.

Procedure

Commence by heating some distilled water in a beaker. Then weigh out (or use a calibrated pipette) 1·0 g of prepared fat (not the original butter or margarine) into flask A and add 4 ml of glycerol and 0·4 ml of sodium hydroxide solution (50% by weight). Hold the flask by using several thicknesses of paper or a test-tube holder and heat it gently over (not in) a *small* flame, swirling and mixing the liquid continuously until the fat is completely saponified. This is shown by the whole liquid becoming perfectly clear. Pour 20 ml of the boiling water (from the beaker, see above) into a 25-ml measuring cylinder and add it (*drop by drop* at first) to flask A, whilst holding it against a cold tile to partially cool it. Swirl the flask continuously after each addition. Then add 10 ml of 7 N sulphuric acid, swirl the mixture and confirm that the liquid is acidic by the addition of a tiny piece of litmus paper (if not acidic, add more acid). Add small glass beads to A, connect it to the apparatus and distil the liquid into B so that 22 ml is collected in 10–12 min. Remove B and place a small beaker, D, under the condenser receiver. Filter the mixed distillate through a 7-cm filter-paper, F, into a test-tube and titrate 10 ml of filtrate with 0·02 N NaOH solution using phenolphthalein as indicator (R ml).

$$\text{Reichert value} = 1 \cdot 1 \ (2R - \text{blank})$$

For the Polenske value, pour 3 ml of cold water successively through (*a*) the condenser into beaker D, (*b*) 22-ml flask B and (*c*) filter-paper F. Repeat these washings with a further 3 ml of water. Dissolve the washed water-insoluble acids by passing three 3-ml portions of neutral alcohol through (*a*), (*b*) and (*c*), collecting the solution in a small flask. Add phenolphthalein and titrate the solution with 0·02 N NaOH (P ml).

$$\text{Polenske value} = P - \text{blank}$$

Genuine butter fat usually gives a Reichert value of 24–32 corresponding to Polenske values of 1·7–3·5. Most other oils and fats have Reichert values not greater than 6, but coconut oil and palm kernel oil give Polenske values of about 16 and 10, respectively. These values are therefore very useful for assessing the proportion of butter fat present in mixtures such as those extracted from butter confectionery and canned cream soups[13, 32].

The *Kirschner value* is the volume of alkali required to neutralise the water-soluble volatile fatty acids that form water-soluble silver salts distilled from the fat. In the method, silver sulphate is shaken with the liquid obtained in titrating the Reichert value. After filtering, the filtrate is transferred to the distillation flask, sulphuric acid and aluminium wire are added, and the liquid is distilled and titrated as for the Reichert value[13, 31]. Genuine butter fat gives Kirschner values from 20·5 to 26·4[13, 32].

MARGARINE

Margarine is made by mixing melted fat with cultured skim milk, salt, emulsifier, colour and vitamins A and D. It is cooled and kneaded into a homogeneous product[33, 34]. Although traditionally hydrogenated fish and whale oils were often used, most modern products contain blends of vegetable oils derived from coconut, palm kernel, maize and arachis oils.

The Margarine Regulations 1967[35] require all margarine sold by retail to contain at least 80% of fat of which *not more* than one-tenth by weight may be milk fat. The maximum water content is 16%. Also, to make it approximately nutritionally equal to butter, margarine must be fortified by the addition of 760–940 i.u. of vitamin A and 80–100 i.u. of vitamin D.

The examination of margarine follows closely the procedure suggested for butter (p. 152), the main determinations being those for water and salt. A method for vitamin A was formerly prescribed in statutory regulations[13, 36]. Industrially, much attention is paid to physical properties such as plasticity, spreadability, consistency and the melting point[34].

CHEESE

Most types of cheese are produced from curd. Souring milk is clotted with rennet, the coagulum is divided and the whey separated by cutting, heating and acid development. The milled curd is then salted and pressed. Differences in composition and properties arise due to varying production techniques, particularly during the ripening, which may last from a few days to a year[37–40]. Apart from production skills, the main factors that control cheese quality are the hygienic condition of the milk and the starter used[41].

Statutory standards[42] prescribe standards for milk fat (on the whole cheese or in the dry matter) and maxima for water, of which examples are given in *Table 6.2*, together with various labelling provisions.

Table 6.2. EXAMPLES OF STANDARDS PRESCRIBED IN THE CHEESE REGULATIONS[42]

Variety of cheese	Milk fat, % (minimum)		Water, % (maximum)
	On whole cheese	On dry matter	
Cheddar	—	48	39
Cheshire, Gloucester	—	48	44
Caerphilly, Wensleydale, Stilton	—	48	46
Lancashire	—	48	48
Full fat hard cheese	—	48	48
Full fat soft cheese	20	—	60
Cream cheese	45	—	—
Full fat processed cheese (except Cheddar)	—	48	48
Processed Cheddar cheese	—	48	43
Cheese spread	20	—	60

Reference methods of analysis are described in a British Standard[43] for moisture (dry at 100–102 °C), fat (by acid digestion—see *Method 6.20*), salt, N and pH. Examination of the raw milk used and various forms of process control are discussed by Davis[41]. The main industrial tests on the product are those for moisture (dry at 100 °C), fat (by the Gerber method—see below), salt, acidity or pH (direct or after softening with water). According to the consistency of the cheese, samples for analysis are prepared by grating them in a domestic grater, cutting with a sharp knife or grinding in a mortar. The prepared sample must be immediately transferred to a stoppered jar and weighed out as soon as possible for analysis.

METHOD 6.19. ESTIMATION OF FAT IN CHEESE BY THE GERBER METHOD[9, 10]

Apparatus—See also *Method 6.3* (p. 132).

Cheese butyrometer tubes Graduated up to 40%, and having suitable stoppers. See also Note (*c*).

Funnel For weighing cheese (*Figure 6.6*).

Camel-hair brush.

Reagents—See *Method 6.3* and Note (*d*).

Procedure

Counterbalance the weighing funnel plus stopper (*Figure 6.6*) and weigh into it 3 ± 0.001 g of sample (*a*).

Pipette 10 ml of Gerber sulphuric acid into the cheese butyrometer and add sufficient warm water (40 °C) to form a layer about 6 mm deep on top of the acid. Insert the funnel plus sample in the cheese butyrometer. Withdraw the stopper and transfer all the cheese to the butyrometer with the aid of the camel-hair brush. Add 1 ml of amyl alcohol by pipette followed by more warm water until the butyrometer is filled to about 5 mm below the shoulder. Insert the stopper and mix the contents of the bulb of the tube before inverting it. Continue shaking and inverting the tube until all the sample is digested and then place it in water at 65 °C for 3–10 min. Centrifuge the tube for 5–6 min at 1100 rpm, return it to the bath for 3–10 min and read off the fat percentage to the nearest 0.3% (one-third of the smallest scale division) (*b*). From the moisture value, calculate the fat content in the dry matter:

$$\text{Fat in dry matter } (\%) = \frac{\% \text{ fat}}{100 - \% \text{ moisture}} \times 100$$

This value is related to the amount of fat in the milk used (see also *Table 6.1*).

Notes

(*a*) With soft cheese, it is preferable to weigh out the sample in a small dish and transfer it to the acid-containing butyrometer with the aid of a few millilitres of hot water. For cheese containing over 40% of fat, use 1.5 g of sample and double the final reading.

(*b*) Re-centrifuging is advisable. See also Note (*a*).

(*c*) The determination can also be carried out in a cream butyrometer previously marked with the level to which the tube should be finally topped up with water. A 5-g amount of sample is weighed directly into the tube and a hot mixture of equal volumes of Gerber acid and water is added and then the amyl alcohol.

(*d*) The method can also be performed with Macdonald's reagent (see *Method 6.23*).

METHOD 6.20. ESTIMATION OF FAT IN CHEESE BY THE WERNER–SCHMID PROCESS (MODIFIED FROM BS 770[43])

Principle of method

The sample is heated with hydrochloric acid, which dissolves the casein, and the liberated fat is extracted with mixed ethers. The method is a general one for foods that contain relatively little sugar.

Procedure

Weigh out 1–2 g of prepared sample (containing 0·3–0·7 g of fat) into a 100-ml lipped beaker, moisten the sample with a few drops of ammonia of sp. gr. 0·880 and macerate it thoroughly using a flat-ended rod. Add 3 ml of water and 7 ml of concentrated HCl, cover the beaker with a watch-glass and heat it carefully, stirring *frequently* with the rod. After dissolving all the particles by gentle simmering, partially cool the solution, add 10 ml of alcohol and stir. Transfer the solution *carefully* down the rod into a separating funnel, rinsing the beaker and rod with several small volumes of diethyl ether to a total ether volume of 25 ml. Shake the separator vigorously for 1 min, then add 25 ml of light petroleum and shake again. After the layers have separated, run the lower layer into a second separator and repeat the extraction. Carry out the extraction at least three times with the mixed ethers, wash the combined extracts with a small volume of water and filter into a suitable flask, washing the filter with a small volume of diethyl ether. Then remove the solvent and weigh the extracted fat as described in *Method 6.4*. For the calculation of the fat in the dry matter see p. 158.

METHOD 6.21. ESTIMATION OF SALT IN CHEESE

Reagent

Ferric alum indicator Dissolve 50 g of ferric ammonium sulphate in a mixture of 95 ml of water and 5 ml of 5 N nitric acid.

Procedure

Weigh 2 g of prepared sample into a 250-ml conical flask and add 10 ml of water and 25 ml of 0·05 N $AgNO_3$. Warm the mixture to 80 °C, swirl it vigorously to disperse the sample, add 10 ml of concentrated HNO_3 and digest the curd by gentle boiling for about

10 min. At this stage the silver chloride should be granular, the liquid a clear lemon colour and the fat layer free from solid. Add 0·3 g of urea to the hot solution, mix, cool, add 1 ml of nitrobenzene and mix again. Add 2 ml of indicator and 50 ml of water and titrate the excess silver nitrate by titration with 0·05 N KCNS until an orange tint persists for 15 s.

$$1 \text{ ml of } 0·05 \text{ N KCNS or AgNO}_3 \equiv 0·00292 \text{ g of NaCl}$$

METHOD 6.22. ESTIMATION OF THE ACIDITY OF CHEESE

Add warm water (40 °C) to 10 g of sample to produce a total volume of 105 ml. Shake the mixture vigorously, filter it and titrate 25-ml portions ($\equiv 2·5$ g of sample) with 0·1 N NaOH using phenolphthalein as indicator. Calculate the acidity as lactic acid.

$$1 \text{ ml of } 0·1 \text{ N NaOH} \equiv 0·009 \text{ g of lactic acid}$$

For the determination of the pH of soft cheese, the electrodes can be pushed directly into the sample. Hard cheeses should be softened first by mixing them with the minimum volume of distilled water.

ICE CREAM

Ice cream is manufactured from milk, fat, sugar, emulsifier, stabiliser, flavour and colour[44]. With 'dairy ice cream' and 'milk ice' all the fat present must be milk fat. Aspects of the processing (ingredients, mixing, homogenising, pasteurising, freezing and packaging), chemical analysis, microbiology and properties of ice cream have been reviewed by Rothwell[45].

The main statutory compositional standards for ice cream prescribed in The Ice-Cream Regulations 1967[46] are given in *Table 6.3*. Alternative standards are prescribed for products containing fruit. The Ice-Cream (Heat Treatment) Regulations prescribe minimum pasteurisation temperatures (150–175 °F for varying times) or sterilisation at not less than 300°F for at least 2 s followed by storage at a temperature not exceeding 28°F[47]. Water ices and ice lollies with a pH of 4·5 or less are, however, exempt from such provisions.

Reference methods of analysis are described in a British Standard[48] for total solids (dry on sand at 100°C), fat (Rose–Gottlieb method using 4 g of sample, 1·5 ml of ammonia of sp. gr. 0·880, 7 ml of water and heat to 65 °C for 15 min before adding 10 ml of alcohol and extracting the fat with mixed ethers), sugars, N, casein, ash,

Table 6.3. THE MAIN STATUTORY COMPOSITIONAL
STANDARDS FOR ICE CREAM[46]

| Type of ice-cream | Minimum statutory standards | |
	Milk solids not fat, %	Fat, %
Ice-cream (general)	7·5	5
Milk ice	7	2·5
Parev Kosher ice	—	10

Parev Kosher ice must not contain milk fat or any other milk derivative.

calcium and phosphorus. Methods for the colony count, coliforms and the methylene blue test are described in a further British Standard[49]. Apart from microbiological tests, the main industrial tests on the product are those for total solids, fat (Gerber or Macdonald method), sugars, milk solids other than fat, and over-run. Sugars are determined by Lane and Eynon's method on a 4% cleared solution before inversion and a 1% solution after inversion. In preparing samples for analysis, the ice cream is transferred to a stoppered bottle, which is placed in hot water at a temperature not exceeding 45 °C. Shake the melted sample and cool it to room temperature within 15 min. All analyses should be commenced as soon as possible after preparing the sample. Over-run must be estimated on the original sample.

METHOD 6.23. ESTIMATION OF FAT IN ICE CREAM USING MACDONALD'S REAGENT IN A GERBER BUTYROMETER

Numerous workers have proposed the use of neutral and alkaline reagents, which are less corrosive than sulphuric acid, for the rapid estimation of fat in dairy products. Macdonald[50] has developed a neutral reagent that is used in a Gerber milk butyrometer. The method has been applied to milk and other dairy products. The method described below for ice cream can also be applied to condensed milk.

Macdonald's reagent Dissolve 1·1 g of Tween 85 (Honeywill–Atlas) in 25 ml of alcohol (industrial methylated spirit, 66 OP) and add 3 ml of n-butanol. Dissolve 1 g of EDTA (disodium salt) in the minimum volume of hot water, neutralise the solution with 1 N NaOH to phenolphthalein and dissolve in it 5 g of trisodium citrate and 5 g of sodium salicylate. Add the latter mixture to the Tween 85 solution and dilute the solution with water to 100 ml. Invert before use.

161

Procedure

Weigh out $4\pm0\cdot01$ g of sample in a small dish. Add 3 ml of hot water, stir the mixture and transfer it down the rod into a Gerber milk butyrometer containing 10 ml of Macdonald's reagent. Rinse out the dish with two 2·5-ml portions of hot water. Stopper the butyrometer, invert it at least twice to mix the contents and place it in a water-bath at 65 °C for 5 min. Remove the tube from the water, shake it vigorously and return it to the bath for 5 min. Then shake it again and centrifuge the butyrometer at 1100 rpm for 5 min and replace it in the water-bath for 5 min and take a reading. Re-centrifuge and take a further reading.

$$\text{Fat } (\%) = \text{Milk butyrometer reading} \times 2\cdot87$$

An amount of sample of 4 g is also used with Gerber sulphuric acid in a milk butyrometer.

METHOD 6.24. ASSESSMENT OF MILK SOLIDS OTHER THAN FAT (MSNF) IN ICE CREAM USING THE FORMOL TITRATION[51]

Exactly neutralise 10 g of ice cream to phenolphthalein with 0·1 N NaOH. Then add exactly 3 ml of formalin and neutralise again with 0·1 N NaOH (*a* ml). Also titrate 3 ml of formalin with 0·1 N NaOH to phenolphthalein (*b* ml)

$$\text{MSNF } (\%) = 5\cdot67 \, (a-b)$$

A potentiometric modification has been described by Hill and Stone[52].

The MSNF can also be assessed from the calcium figure (*Method 4.3*). The dry fat-free solids of milk contains $1\cdot95\%$ of CaO.

METHOD 6.25. ESTIMATION OF THE OVER-RUN OF ICE CREAM

Over-run is the percentage increase in volume caused by whipping air into ice cream in order to produce a more delicate texture. Mathematically:

$$\text{Over-run } (\%) = \frac{(y-x)}{y} \times 100 =$$

$$\frac{\text{Volume of ice cream} - \text{volume of mix}}{\text{Volume of mix}} \times 100$$

where x = sp. gr. of solid ice cream before melting and y = sp. gr. of the remelted ice cream after removal of air.

The following method for over-run is suitable for routine work[53]. A sample is normally removed prior to filling and examined near the plant.

Apparatus

Cup Plastic or metal cup of a capacity such that it holds, when full, about 180–420 g of ice cream. The depth should not exceed the diameter and the handle should not extend beyond the top of the cup to facilitate striking off excess of sample.
Striker A flat-metal straight edge at least 2 cm wide.
Semi-accurate balance To give rapid weighings to the nearest gram.

Procedure

Weigh the empty cup (W_1). Then fill the cup with ice cream, mix, strike off the excess cleanly across the top and weigh (W_2), so that the net weight of mix is $W_2 - W_1$. Calculate the weight of the cup full of ice cream at each over-run value over the desired range (e.g., 60–100%) by means of the following equation:

$$\left(\frac{\text{Net wt. of cup of mix}}{100 + \text{desired over-run}} \times 100 \right) + \text{wt. of empty cup} =$$

$$\text{wt. of ice cream and cup}$$

If the net weight is 404 g, the empty cup weighs 78 g and the desired over-run is 80%, then:

$$\left(\frac{404}{100 + 80} \times 100 \right) + 78 =$$

$$302\ g = \text{wt. of ice cream and cup at } 80\% \text{ over-run}$$

Then construct a table relating 'weight of cup + ice cream' in grams and the 'over-run percentage' covering, for example, 60, 65, 70, 75, 77, 79, 80, 81, 83, 85, 90, 95 and 100%. Then, in use, fill the cup with the sample, strike off the excess, weigh and obtain the over-run from the table (cf., AOAC).

DAIRY PRODUCTS

REFERENCES

1. DAVIS, J. G., *Milk Testing*, 2nd edn, Dairy Industries Ltd, 9 Gough Sq, London, E.C.4 (1959)
2. *Method for Determination of the Freezing-point Depression of Milk (Hortvet Method)*, BS 3095:1959, British Standards Institution, London
3. *The Milk (Special Designation) Regulations 1963 (SI 1963 No. 1571; 1965 No. 1555)*, H.M.S.O., London
4. *The Sale of Milk Regulations 1939 (SR & O 1939 No. 1417)*, H.M.S.O., London
5. DAVIS, J. G. and MACDONALD, F. J., *Richmond's Dairy Chemistry*, 5th edn, Griffin, London (1953)
6. *Methods of Microbiological Examination for Dairy Purposes*, BS 4285:1968, British Standards Institution, London
7. *The Milk (Special Designations) (Scotland) Order 1965 (SI 1965 No. 253) (S. 11)*, H.M.S.O., London and Edinburgh
8. *Methods for Sampling Milk and Milk Products*, BS 809:1963, British Standards Institution, London
9. *Gerber Method for the Determination of Fat in Milk and Milk Products, Part 1: Apparatus*, BS 696:1955, British Standards Institution, London
10. *Gerber Method for the Determination of Fat in Milk and Milk Products, Part 2: Methods*, BS 696:1969, British Standards Institution, London
11. *Chemical Analysis of Liquid Milk and Cream*, BS 1741:1963, British Standards Institution, London
12. *Density Hydrometers for Use in Milk*, BS 734:1959, British Standards Institution, London
13. PEARSON, D., *The Chemical Analysis of Foods*, 6th edn, Churchill, London (1970)
14. THOMAS, S. B., JONES, T. I. and GRIFFITHS, D. G., *Dairy Ind.*, **28**, 812 (1963)
15. THOMAS, S. B. and MAKINSON, P. E., *Dairy Ind.*, **29**, 432 (1964)
16. FISCHER, B. M., *J. Soc. Dairy Technol.*, **18**, 230 (1965)
17. FELLOWS, P. V., *Dairy Ind.*, **18**, 309 (1953)
18. CHAPMAN, R. A. and MCFARLANE, W. D., *Can. J. Res.*, **23B**, 91 (1945)
19. CROWE, L. K., JENNESS, R. and COULTER, S. T., *J. Dairy Sci.*, **31**, 595 (1948)
20. ROTHWELL, J., *Process Biochem.*, **3**, No. 1, 19 (1968)
21. *The Cream Regulations 1970 (SI 1970 No. 752)*, H.M.S.O., London
22. *The Condensed Milk Regulations 1959 (SI 1959 No. 1098)*, H.M.S.O., London
23. *Methods for the Chemical Analysis of Condensed Milk*, BS 1742:1951, British Standards Institution, London
24. KING, N., *Dairy Sci. Abstr.*, **28**, No. 3, 105 (1966)
25. *The Dried Milk Regulations 1965 (SI 1965 No. 363)*, H.M.S.O., London
26. *Methods for the Chemical Analysis of Dried Milk*, BS 1743:1968, British Standards Institution, London
27. MAUGHAN, J. H. D., *Dairy Ind.*, **33**, 627 (1968)
28. *The American Dry Milk Institute Bulletin*, No. 916, Chicago, Illinois, 1963 (revised edn)
29. ROTHWELL, J., *Process Biochem.*, **1**, No. 4, 207 (1966)
30. *The Butter Regulations 1966 (SI 1966 No. 1074)*, H.M.S.O., London
31. *Methods for the Chemical Analysis of Butter*, BS 769:1961, British Standards Institution, London
32. WILLIAMS, K. A., *Analyst, Lond.*, **74**, 508 (1949)
33. ANDERSEN, A. J. C. and WILLIAMS, P. N., *Margarine*, 2nd edn, Pergamon Press, Oxford (1965)
34. VAN DER VET, A. P., 'Edible Fats and Oils', in HERSCHDOERFER, S. M. (Editor), *Quality Control in the Food Industry*, Vol. 2, Academic Press, London and New York, 394 (1968)

35. *The Margarine Regulations 1967 (SI 1967 No. 1867)*, H.M.S.O., London
36. *The Food Standards (Margarine) Order 1954 (SR & O 1954 No. 613)*, H.M.S.O., London
37. DAVIS, J. G., *Cheese*, Vol. I, Churchill, London (1965)
38. DAVIS, J. G., *Cheese*, Vol. II, Churchill, London (1965)
39. MINISTRY OF AGRICULTURE, 'Cheesemaking', *Bulletin No. 43*, H.M.S.O., London (1959)
40. SANDERS, G. P., *Cheese Varieties*, U.S. Dept. of Agriculture, Washington, D.C., U.S.A. (1953)
41. DAVIS, J. G., 'Dairy Products', in HERSCHDOERFER, S. M. (Editor) *Quality Control in the Food Industry*, Vol. 2, Academic Press, London and New York, 127 (1968)
42. *The Cheese Regulations 1970 (SI 1970 No. 94)*, H.M.S.O., London
43. *Methods for the Chemical Analysis of Cheese*, BS 770:1963, British Standards Institution, London
44. ROTHWELL, J., *Dairy and Ice Cream Industries Directory*, United Trade Press, London, 120 pp. (1968)
45. ROTHWELL, J., *Dairy Sci. Abstr.*, **22**, No. 10, 483 (1960)
46. *The Ice-Cream Regulations 1967 (SI 1967 No. 1866)*, H.M.S.O., London
47. *The Ice-Cream (Heat Treatment, etc.) Regulations 1959 (SI 1959 No. 734; 1963 No. 1083)*, H.M.S.O., London
48. *Methods for the Chemical Analysis of Ice Cream*, BS 2472:1966, British Standards Institution, London
49. *Methods of Microbiological Examination of Milk Products*, BS 4285:1968, Suppl. No. 1 (1970), British Standards Institution, London
50. MACDONALD, F. J., *Analyst, Lond.*, **84**, 287 (1959)
51. CROWHURST, B., *Analyst, Lond.*, **81**, 123 (1956)
52. HILL, R. L. and STONE, W. K., *J. Dairy Sci.*, **47**, 1014 (1964)
53. ENGLAND, C. W., *Ice Cream Rev.*, **35**, No. 11, 40 (1952)

7

FLESH FOODS—MEAT AND FISH

Assessment of Freshness—Assessment of Meat Content—Brines—Cured Meat—Smoked Fish—Mercury in Canned Tuna

MEAT

Apart from the assessment of the proportion of fat present, the examination of raw meat is not commonly carried out in the laboratory. When the condition is in doubt other tests may be called for, such as checks on rancidity. *Post-mortem* examination of carcases, cattle, sheep and pigs for disease, etc., is normally carried out by experienced inspectors[1]. In general, an attractive red colour with marbling fat is desirable in lean meat and the fat should be white rather than yellow[2]. For products, the ability to hold water (binding) varies considerably, together with composition and other properties, according to the cut.

FISH

The identification of species[3] and decisions as to freshness of fish are more usually the prerogative of the inspector rather than the laboratory. The laboratory may, however, have to give an opinion in borderline cases of spoilage.

In distinguishing one species of fish from another, the following should be considered:

(*a*) Length, shape (round, flat, etc.) and colour.

(*b*) Distinguished markings.

(*c*) The number and position of the fins, especially the dorsal fins.

(*d*) The shape, colour and prominence of the lateral line.

Some characteristics of fresh and spoiled white fish are given in *Table 7.1*.

Table 7.1. SOME CHARACTERISTICS OF FRESH AND SPOILED RAW FISH

Characteristic	Fresh fish	Spoiled fish
Odour	Fresh or seaweedy	Putrid
Surface	Glossy, thin, transparent slime	Dull, thick, clotted brown slime
Texture	Firm	Soft
Backbone	No discoloration and flesh difficult to strip from backbone	Red discoloration along backbone. Flesh easy to strip off
Gills	Red, no slime	Grey or yellowish, very slimy
Eyes	Prominent. Pupil jet black. Cornea transparent	Shrunken. Pupil milky. Cornea opaque

METHOD 7.1. SAMPLING OF FLESH FOODS AND PRODUCTS

As flesh foods and their products tend to be highly heterogeneous, it is often difficult to prepare a truly representative sample for analysis. Great care must therefore be taken in preparing the sample.

Procedure

With meat, take as representative a sample as possible. With fish, cut off the head and remove the skin and bones. With sausage, remove the skin. Pass the flesh, sausage or other product through a mincer and then mix it thoroughly in a pestle and mortar. Put the mixed sample back in the mincer and repeat the operation. With meat, the mincing and mixing must be carried out at least twice. Take care to re-incorporate any connective tissue or juice that separates during mincing. Finally, transfer the sample to a labelled and stoppered jar and store it in a refrigerator.

ASSESSMENT OF SPOILAGE OF FLESH FOODS BY CHEMICAL METHODS

Although lengthy microbiological techniques must be used to confirm the absence of pathogens, relatively rapid chemical methods are useful for assessing the degree of spoilage or acceptability of flesh foods. In the case of beef and white fish, the results of such methods have been correlated with organoleptic findings (*Table 7.2*). Such values are applicable to raw (not cooked) flesh. Although spoilage of meat and fish is very complex, the methods included in this Section can be considered on the basis of one or more of the following reactions, the applicability of which to particular types of flesh is stated in *Table 7.3*.

(1) Breakdown (deamination) of protein producing ammonia, indole, skatole, H_2S, etc.

(2) Formation of trimethylamine (TMA) from trimethylamine oxide (TMO) due to reduction by bacteria:

$$CH_3CHOH \cdot COOH + 2(CH_3)_3NO \rightarrow$$

Lactic acid   TMO

$$CH_3 \cdot COOH + CO_2 + H_2O + 2(CH_3)_3N$$

Acetic acid   TMA

(3) Formation of dimethylamine (DMA) (pre-cursor unknown).

(4) Formation of ammonia from urea due to bacterial action:

$$CO(NH_2)_2 + H_2O \rightarrow 2NH_3 + CO_2$$

Urea

(5) Spoilage of fat causing hydrolysis (FFA production) and oxidative and other forms of rancidity.

For analytical convenience, the volatile bases (TVB or TVN ≡ ammonia + amines) are often estimated as a group.

Table 7.2. ASSESSMENT OF ACCEPTABILITY OF BEEF[4] AND WHITE FISH[5] FROM TOTAL VOLATILE NITROGEN (TVN) VALUES *(Method 7.2)*

Flesh	Acceptability or condition	Total volatile nitrogen, mg N/100g flesh
Beef	Fresh (mean)	13
Beef	Acceptable	⩽17
White fish	Fresh	⩽20
White fish	Acceptable	20–30
White fish	Spoiling	>30
White fish	Spoiled	⩾50

N.B. The above values apply to raw flesh. Cooked and processed products usually give much higher values.

In general, therefore, the freshness of white fish can be assessed from the TVB or TMA values and that of fatty fish and meat from consideration of both the TVN (mainly ammonia) and the fat rancidity values.

Table 7.3. APPLICABILITY OF THE VARIOUS SPOILAGE REACTIONS TO
THE DIFFERENT TYPES OF FISH AND MEAT

Flesh	Reaction number (see p. 168)				
	1	2	3	4	5
Meat	Yes	No	No	No	Yes
White fish	Yes	Yes	Yes	No	No
Fatty fish	Yes	Yes	Yes	No	Yes
Freshwater fish	Yes	No(a)	No(a)	No	Yes
Elasmobranch	Yes	Yes	Yes(b)	Yes	No

Notes

    (a) Volatile amines are produced in some types of salmon.

    (b) No dimethylamine is formed in dogfish.

    It should be borne in mind that the reaction that first affects the organoleptic properties varies with the type of flesh:

        (i)   In meat, protein breakdown is usually noticeable before fat spoilage.

        (ii) With white fish, the first signs of staleness are associated with TMA formation.

        (iii) With fatty fish, rancidity precedes production of TVN.

METHOD 7.2. ESTIMATION OF THE TVN BY LUCKE AND GEIDEL'S MACRO DISTILLATION METHOD

Principle of method

The TVN is released by boiling the flesh directly with magnesium oxide, which also prevents volatile acids from distilling over into the boric acid. The distillate is titrated with standard acid. Some TVN is produced from the protein, but as the rate of boiling and time of distillation are standardised, the results can be interpreted on a comparative basis.

Apparatus and Reagents—See Method 2.6a (p. 52).

Procedure

Macerate 10 g of comminuted sample with 50 ml of fresh tap water in a small mechanical macerator. Wash the macerate into the distillation flask of the macro apparatus (*Figure 2.11*, p. 52) with 250 ml of fresh tap water and add 1–2 g of MgO. To the receiving flask add 25 ml of 2% boric acid solution and screened methyl red indicator. Connect the apparatus, the receiver tube dipping below the surface of the boric acid solution. Heat the liquid so that it boils in exactly 10 min and continue heating the distillation flask *at the same rate* throughout the distillation. Distil the liquid over for *exactly* 25 min. Wash down the condenser and tube with distilled water and titrate the volatile nitrogen with 0·1 N sulphuric acid.

169

$$\text{TVN} = \text{Titration (ml of 0·1 N acid)} \times 14$$
(mg N/100 g)

Table 7.2 gives the suggested limits of acceptability of meat and fish that have been suggested following correlation with organoleptic considerations[4, 5]. This suggests that meat is acceptable up to a TVN value of about 17 mg N per 100 g with a corresponding value for white fish of about 30. Most consumers are conditioned, however, to eating fish that is in a more advanced state of spoilage than is handled by research workers and higher values might be acceptable. It is advisable, however, for the processor of products to use fish in as fresh a state as possible.

Pearson and Muslemuddin[6-9] have shown that by carrying out the distillation under reduced pressure there is no breakdown of protein during the distillation.

METHOD 7.3. ESTIMATION OF TVN AND TMA BY THE CONWAY MICRODIFFUSION TECHNIQUE

Principle of method

On adding potassium carbonate solution to fish juice, volatile nitrogen is released, which diffuses into boric acid solution. The absorbed base is then titrated with acid. By using deproteinated juice and carrying out the diffusion below, say, 45 °C, no additional breakdown of protein is possible during the distillation (cf. *Method 7.2*).

If formalin is added to the juice prior to the addition of alkali, the ammonia reacts to form hexamethylenetetramine and only the TMA diffuses over into the boric acid:

$$6HCHO + 4NH_3 \rightarrow (CH_2)_6N_4 + 6H_2O$$

Formalin   Ammonia   Hexamethylenetetramine

Apparatus

Conway microdiffusion dish with lid (Figure 7.1) Prior to use, soak the dishes and lids in chromic acid mixture overnight, then in distilled water, and dry them in a 100 °C oven. Also, grease the rim of the dish and the lid.

Microburette Total capacity 2 ml.

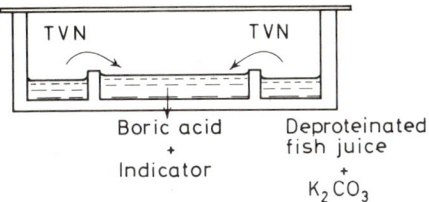

Figure 7.1. Conway microdiffusion method for estimation of volatile bases:
Potassium carbonate is added to the deproteinated fish juice in the outer
compartment of the Conway unit. The volatile nitrogen (TVN) diffuses over into
the boric acid solution contained in the central compartment

Reagent

Conway boric acid Stir 100 ml of industrial alcohol with 5 g of boric
acid and add 350 ml of water. When the acid has dissolved, add 5 ml
of indicator (0·066% methyl red and 0·033% bromocresol green–
alcohol). Then add alkali to produce a faint reddish colour and
make up the volume to 500 ml with alcohol.

Preparation of extract

Mash thoroughly 2·5 g of trichloroacetic acid with 50 g of minced
fish (or meat) in a porcelain basin using a kitchen fork. Allow the
mixture to stand for 30 min, then filter it, first on a Buchner funnel
(press the sample well down) and then using an ordinary funnel
(Whatman No. 5 filter-paper). Store the filtered extract at 0 °C in a
small screw-capped bottle.

N.B. Determinations should be carried out in duplicate together
with a blank.

Determination of total volatile nitrogen (TVN)

Place 2 ml of Conway boric acid solution into the centre compartment
of a Conway microdiffusion unit. Pipette 1 ml of extract if the fish
is fresh (or a suitable aliquot from a 2/10 ml dilution if spoiled) on
one side of the outer compartment. Grease the waxed Conway lid
(rough side) and place it over the unit so that the opposite side of the
outer compartment is uncovered. Pipette 1 ml of saturated filtered
potassium carbonate solution into the outer compartment through
this gap and quickly slide the lid across the edge to form an airtight

seal. Rotate the unit gently (a special shaker is available) and allow it to stand undisturbed overnight. In the morning, remove the cover and titrate the bases, which have diffused over, with 0·0143 N (N/70) sulphuric acid. Express the result as mg N/100 ml of extract.

1 ml of 0·0143 N (N/70) sulphuric acid ≡ 0·2 mg N

Determination of trimethylamine (TMA)

Proceed as for TVN but with the addition of 20 drops of the neutral formalin to the outer compartment space evenly round the circle after the addition of the extract.

It should be borne in mind in interpreting results that white fish contain about 80 % of water, so that results calculated on 100 ml of juice are approximately 25 % higher than those based on 100 g of muscle as obtained in *Method 7.2*.

A simple colorimetric method for estimating ammonia in fish, in which the reaction with bromine and thymol is used, has been described by Fernandez-Flores and Salwin[10].

METHOD 7.4. ASSESSMENT OF FAT SPOILAGE AND RANCIDITY

Fat rancidity methods are applied to meat, fatty fish and freshwater fish. Fat spoilage can be readily assessed by estimating the FFA and peroxide value on a common chloroform extract, as described in *Method 5.8*. Although the interpretation of such values presents difficulties, the following values have been suggested as suitable limits of acceptability (based on the extracted fat):

Beef—1·2 % FFA (as oleic acid)[11].

Fatty fish—peroxide value of 5–10 mequiv./kg.

Canned herring—acid value of 2·75[12].

METHOD 7.5. DETERMINATION OF THE THIOBARBITURIC ACID (TBA) NUMBER[13]

Principle of method

Thiobarbituric acid (TBA) is heated with a distillate obtained from the acidified sample. A red pigment is formed between TBA and the oxidised lipids so that the colour intensity obtained increases as oxidative rancidity advances. There is evidence that malonaldehyde is responsible for the reaction with TBA[14].

172

Thiobarbituric acid reagent (TBA) Dissolve 0·2883 g of thiobarbituric acid in 90% acetic acid by slight warming and make the volume up to 100 ml with 90% acetic acid.

Procedure

Blend 10 g of meat or a fish with 50 ml of water in a mechanical blender and wash the mixture into a suitable distillation flask with 47·5 ml of water. Add 2·5 ml of 4 N HCl to bring the pH to 1·5. Add an antifoam preparation and antibumping granules and connect the flask to the distillation apparatus. Heat the flask by means of an electric mantle so that 50 ml of distillate is collected in 10 min from the time boiling commences. Pipette 5 ml of distillate into a stoppered tube, add 5 ml of TBA reagent, stopper the tube, shake it and place it in boiling water for exactly 35 min. Prepare a blank at the same time using 5 ml of water plus 5 ml of reagent. Cool the tubes in water for 10 min and measure the optical density against the blank at 538 nm.

TBA number = 7·8 × optical density (mg malonaldehyde/kg)

Partly as a result of variability in fresh samples, the interpretation of TBA values in relation to acceptability limits appears to present difficulties.

BASIC COMPOSITION OF MEAT

Basically, meat consists of a mixture of lean tissue and outside fat. The following can be assumed to be fairly typical percentage composition figures for pieces of beef, mutton and pork containing a reasonable proportion of fat:

	Water	*Fat*	*Protein* $N \times 6·25$	*Ash*
Beef	59 (76·7)	23	17 (22·1)	1
Mutton	55 (77·4)	29	15 (21·1)	1
Pork	46 (76·7)	40	13 (21·7)	1

The figures in parentheses represent the same mean percentages calculated on the fat-free basis—see below.

Very different values may be obtained, however, with other cuts according to the proportion of fat present. This variation can be readily seen by calculating the fat/protein ratio, which for pork may

173

vary from 0·15 to over 100. The composition of fat-free (FF) meat is, however, approximately constant. Taking, for instance, the values for pork above:

$$\text{Protein in fat-free} = \frac{13}{100-40} \times 100 = 21\cdot7\% \ (\equiv 3\cdot5\% \ \text{N/FF})$$

$$\text{Water in fat-free} = \frac{46}{100-40} \times 100 = 76\cdot7\%$$

For all the main types of fresh meat, the amount of water present calculated on the fat-free basis varies from 75 to 78% and hence it is often said that lean meat consists of three-quarters water. Further, although lean meat contains intramuscular fat and outside fat contains protein, the composition, when calculated on the fat-free basis, is approximately constant. As the nitrogen content is the main value used in the calculation of the meat content of products, it is important to have agreed mean N/FF factors for the main types of flesh. Compromise N/FF factors have been proposed by the Society for Analytical Chemistry (SAC) for various types of flesh (as percentages): beef 3·55 (from a range of 2·96 to 4·53), veal 3·35, pork 3·45 (from a range of 2·8 to 4·2), chicken 3·7, turkey 3·65. Such factors apply to raw flesh with a normal water content. There are, however, no agreed factors (at the time of writing) for processed meats.

One interesting finding by the SAC for pork was the increase in the N/FF from 3·26 to 3·65% with increasing live weights of pigs ranging from 140 to 320 lb. From the SAC data, it would appear that, on the average:

$$\text{N/FF for pork (\%)} = 0\cdot0021 \ W + 2\cdot97$$

where W lb = live weight of pig carcase.

CLASSIFICATION AND BASIC COMPOSITION OF FISH

Most commercial fish, including cod, haddock, plaice, herring and salmon, have a bony skeleton and are referred to as teleosts. The elasmobranch species skate, rays and dogfish (rock salmon), however, have a skeleton composed of cartilage.

Although the proportions of the various amino units in the protein show considerable differences, the basic composition of fat-free fish is not unlike that of meat. Further, the teleosts can readily be classified as follows:

(i) White round demersal fish, such as cod and haddock, containing less than 1% of fat.

(*ii*) White flat demersal fish, such as place and lemon sole, containing less than 4% of fat (except halibut, which may contain up to about 9% of fat).

(*iii*) Fatty pelagic fish, such as herring, which has a variable fat content, which in the autumn may rise to about 28%.

(*iv*) Freshwater fish, such as the red-fleshed salmon, containing up to 14% of fat.

Typical compositional ranges for the main classes are given in *Table 7.4*.

Table 7.4. BASIC COMPOSITIONAL RANGES OF CLASSES OF FISH

Class or species of fish	Water, %	Fat, %	Protein, %	Ash, %
White, round	79–84	0·1–0·9	15–20	1
White, flat	78–82	0·5–4·0	15–19	1
Herring	66 (mean)	7–30	14–20	1
Salmon	67 (mean)	1–14	16–24	1

The SAC have recommended a mean total nitrogen factor for cod of 2·85.

ALLOWANCE FOR NITROGEN PRESENT IN MEAT PRODUCT FILLERS

Allowance for the nitrogen supplied by the fillers used may be necessary when calculating the proportion of meat present in products. For wheat rusk, the filler commonly used in fresh sausage, the SAC recommended that a value of 2·0% should be used as the correction for nitrogen. Cornflour and potato starch ('farina'), however, contain a negligible amount of protein and no correction for nitrogen is then necessary. Of the protein-rich 'fillers' used, skim milk powder contains 6·0% N and soya flour about 6·8%.

ASSESSMENT OF THE MEAT AND FISH CONTENT OF PRODUCTS

TOTAL MEAT CONTENT

In view of the high cost of meat (especially lean meat) and various legal requirements, it is often necessary to control the amount present in products. Some of the main legal requirements are summarised in *Table 7.5*. The method of calculating the meat content is based on the original formula of Stubbs and More, but using the factors later recommended by the SAC. Assuming that the nitrogen in the sample is derived solely from the flesh and cereal filler:

$$\text{Total meat } (\%) = \frac{\% \text{ Total N} - \% \text{ Filler N}}{\text{Mean N/FF for meat}} \times 100 + \% \text{ Extracted fat}$$

$$= \frac{N_T - \% \text{ Filler N}}{N_F} \times 100 + F_{Ext}$$

$$\text{Carbohydrate } (\%) = C =$$
$$100 - (\% \text{ Water} + \% \text{ Fat} + \% \text{ Protein} + \% \text{ Ash}),$$

where $\% \text{ Protein} = \% \text{ Total N} \times 6.25$

$$\text{Filler N } (\%) = K_F C$$

where $K_F = 0.02$ for wheat rusk, $K_F = 0.018$ for pearl barley but $K_F = 0.0$ for maize starch and potato starch.

Therefore,

$$\text{Total meat } (\%) = \left[\frac{N_T - K_F C}{N_F} \times 100 \right] + F_{Ext} \qquad (7.1)$$

The recognised SAC factors for the mean N/FF (N_F) for various meats is given on p. 174, e.g., 3·45 for pork and 3·55 for beef.

Table 7.5. STATUTORY MINIMA PRESCRIBED FOR SOME OF THE MORE IMPORTANT FLESH PRODUCTS

Product	Minimum meat or fish content, %
Pork sausage	65(a)
Beef sausage	50(a)
Meat pie (cooked)	25(a)
Sausage roll (uncooked)	10·5(a)
Sausage roll (cooked)	12·5(a)
Meat paste	70(a)
Fish paste	70(a)
Fish cake	35(a)(c)
Canned meat	95(b)(d)
Cured meat (canned)	90(b)
Luncheon meat (canned)	80(b)

Notes
(a) Calculated as raw meat or fish.
(b) Calculated as raw or after curing or any other similar processing.
(c) FSC have recommended[15] that the minimum should be raised to 40%.
(d) For products declared or implied to contain no food other than meat.

LEAN MEAT CONTENT

For legal purposes, lean meat means 'lean meat free of visible fat'. Meat without any visible fat contains about 4–11 % of intramuscular fat. Outside fat with no visible lean meat contains up to about 93 % of fat, the remainder including water, protein and ash. Assuming that the allowable limit of fat in lean meat is 10 % and that 90 % is the average fat content for outside fat:

$$\text{Lean meat } (\%) = \frac{112 \cdot 5(N_T - K_F C - B F_{Ext})}{N_F} \quad (7.2)$$

where $B = 0 \cdot 003833$ for pork and $B = 0 \cdot 003944$ for beef.
The other symbols are the same as those given above in the calculation of the total meat content.

If the percentage of defatted meat (DM) is equal to F_{Ext}, then the percentage of calculated lean meat (LM) is numerically identical with DM.

If, however, $DM > F_{Ext}$, then $LM > DM$, and vice versa.

SAMPLING—See *Method 7.1*.

METHOD 7.6. ESTIMATION OF WATER IN FLESH PRODUCTS

Obtain a spouted metal dish (approximately 7 cm in diameter) and place a flat-ended rod (about 8 cm long) across the lip (with the flat end in the dish as in *Figure 2.4*, p. 32). Dry the dish and rod in a 100 °C oven for up to 30 min. Transfer the dish and rod to a desiccator, cool and weigh them. Then weigh out accurately about 5 g of the well mixed sample into the dish. Add carefully about 1 ml of industrial methylated spirit and spread the sample evenly across the bottom of the dish, matting it down with the flat end of the rod. Place the dish, rod and contents on the metal part of a water-bath (away from the direct steam; fierce heating might involve losses due to spurting). When the sample appears dry, transfer the dish and contents to the 100 °C oven for 3 h. (While the drying is proceeding, place a clean flask in the 100 °C oven and heat it for about 30 min— for fat determination, see p. 178.) Place the dish and contents in a desiccator, cool and weigh. If necessary, continue the heating for $\frac{1}{2}$-h intervals and weigh to constant weight. Retain the dried solids for the fat estimation (see p. 178).

Calculate the water from the percentage loss in weight.

METHOD 7.7. GRAVIMETRIC ESTIMATION OF FAT USING A SOXHLET OR
BOLTON EXTRACTOR

Transfer the flask (see *Method 7.6* above) to a desiccator and weigh
it as soon as possible after it has cooled. Fit the flask with a Soxhlet
or Bolton extractor (*Figure 2.7*, p. 44) and secure it in a clamp or
stand on the bench. Mould a filter-paper on a large test-tube and
transfer the main bulk of the dry solids (from *Method 7.6* above) into
the paper with some assistance from the rod. Support the paper and
contents in the upper part of the extractor. Add a few millilitres of
light petroleum to the dish and mix the remaining solid thoroughly
with the rod, using the flat end to disintegrate any large lumps. Also,
remove any solid adhering to the rod with a clean spatula and wash
the latter with a further small volume of light petroleum into the
dish. Then transfer the light petroleum in the dish down the rod into
the filter-paper in the extractor. Repeat this washing of the rod and
dish into the extractor with some further *small* volumes of light
petroleum. Plug the top of the paper with de-fatted cotton-wool and
push it down into the lower part of the extractor. Then add more
light petroleum through the top of the extractor, making sure to
wash round the rim to which the paper had previously adhered.
When a Soxhlet extractor is used, the total volume of solvent used
should be such that it is about $1\frac{1}{2}$ times the volume of the extractor
when it is filled to the point where it is just about to siphon over.

After the requisite amount of solvent has been added, attach a
suitable condenser and heat the flask in the apparatus on a special
water-bath or electric mantle. Continue the extraction for 1 h and
then remove the flask with the attached extractor from the bath.
Raise the paper or thimble with tongs and allow the solvent to drain
for a short while into the extractor. Allow the paper and contents
to air-dry and then grind up the solid thoroughly in a mortar. Return
the ground-up solid to the paper, plug it with cotton-wool as before
and continue the extraction for a further period (usually 1–2 h).
After the fat has been completely extracted, remove the flask and
extractor from the bath and take out the paper. Replace the ordinary
extractor with one that is suitable for removing solvent (*Figure
6.3*, page 135). Place it on the apparatus and tap off the condensed
solvent. When the solvent appears to have been removed, detach
the flask, wipe off any water from the outside and place it in a 100 °C
oven for 45 min. Remove the flask from time to time during the
heating and blow air on to the fat by using a hand bellows. Finally,

transfer the flask and fat to a desiccator, cool and weigh it. Retain the extracted fat for estimation of FFA if necessary (*Method 5.5*).

Calculate the fat as a percentage of the original sample (not on the dried solids).

METHOD 7.8. RAPID VOLUMETRIC METHOD FOR THE ESTIMATION OF FAT USING A VAN GULIK BUTRYOMETER

Principle of method

The sample is digested with an acetic–perchloric acid mixture in boiling water and the centrifuged fat is measured in a Van Gulik butyrometer, which is specially designed for use with 5 g of meat product. The sensitivity of the readings is increased by containing much of the fat within the bulb, so that the graduated scale covers only the desired range of readings.

Apparatus

Van Gulik butyrometer (*Astell Laboratory Service Co.*) For use with 5 g of meat product, with graduations at 0 and over the range 20–35%. The main stopper, A, is fitted with a plain (B) or perforated (C) cup. The graduated end of the stem is closed with a small stopper, D (*Figure 7.2*). Tubes can be calibrated to cover any desired range of readings.
Water bath Controlled at 68 °C with a shaking stand for butyrometers.
Gerber centrifuge.

Reagent

Salwin acid reagent 50 ml of glacial acetic acid plus 50 ml of perchloric acid (60%).

Procedure

Weigh $5\pm0\cdot01$ g of sausage into the glass cup (preferably the perforated type, C) mounted in its rubber stopper, A, and insert the stopper tightly into the dry Van Gulik meat butyrometer (*Figure 7.2*). Place the butyrometer in a tall stand (as used for Gerber butyrometers) and add 12–13 ml (*a*) of Salwin acid reagent slowly through the narrow end so that the level is below the zero mark.

179

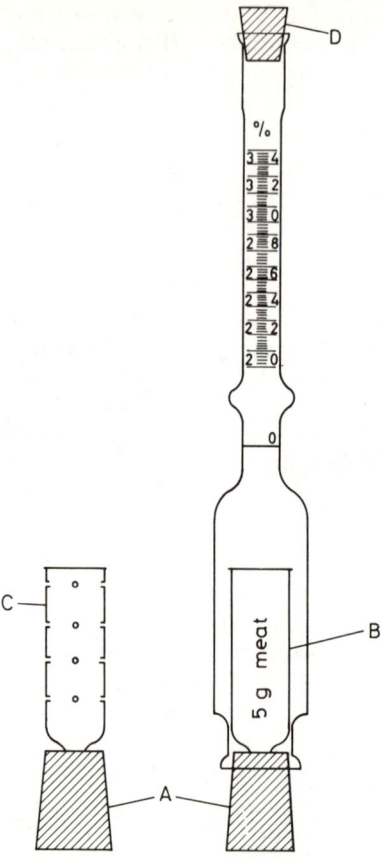

Figure 7.2. Van Gulik meat butyrometer for the volumetric estimation of fat showing plain cup (B) and perforated cup (C) for weighing out sample. (Astell Laboratory Service Co.)

Place the tube in boiling water for about 15–30 min (shaking thoroughly from time to time after inserting the small stopper, D, which should be removed during the heating) until the sample is thoroughly dispersed and digested. Then insert the small stopper, shake the butyrometer, transfer it to a water-bath at 65–70 °C for 5 min and centrifuge it for 3 min at 1100 rpm. Return it to the 65–70 °C bath for 4 min and carefully add more Salwin reagent through the narrow end of the butyrometer so that the top of the lower layer is slightly below the zero mark (*a*). Return the tube to the bath for 2 min and re-centrifuge it for 5 min. Finally, return it

180

to the bath for at least 2 min and read the fat percentage directly from the graduated scale, taking as the level the bottom of the meniscus after adjusting the lower level of the fat column to the zero mark by carefully manipulating the large stopper, A (*b*). After taking the final reading, wash the tube and stoppers thoroughly with hot water and allow it to drain in the Gerber butryometer stand.

Notes

(*a*) After carrying out the method on several samples, the exact total volume of reagent (usually about 14 ml) to be used can be permanently marked on to the butyrometer for future reference. The correct total volume of reagent can usefully be poured into a measuring cylinder prior to the test.

(*b*) *Slight* final adjustments can readily be made by *short* immersion of the *lower* end of the bulb in cold or boiling water according to whether contraction or expansion is necessary.

The above method has the following main advantages over using Gerber sulphuric acid in a Gerber butyrometer:

(*i*) Owing to calibration over the shorter selected range of readings, the fat content can be assessed more accurately.

(*ii*) The digestion with the Salwin reagent is shorter and more thorough than when the sample is treated with 90% sulphuric acid.

(*iii*) A clear fat column is produced, whereas a very dark fat column results after digestion with sulphuric acid, particularly if cereal and spices are present.

Rapid methods for determining fat have been reviewed by Mahmood-ul-Hassan and Pearson[16] and Smith[17]. Procedures based on spectrophotometry[18], the SI-MO-FAT method[19] and the specific gravity[20] of the fat have also been described.

METHOD 7.9. ESTIMATION OF PROTEIN IN FLESH FOODS BY THE KJELDAHL TECHNIQUE

Weigh out 1·8–2·0 g of sample rapidly on to a 9-cm filter-paper resting on a watch- or clock-glass. Then drop the rolled paper plus sample straight into the macro Kjeldahl digestion flask. Estimate the total nitrogen in the sample by the macro Kjeldahl technique (*Method 2.6a*). Use $N \times 6·25$ to convert nitrogen to protein.

METHOD 7.10. TENTATIVE DYE-BINDING METHOD FOR THE ESTIMATION OF PROTEIN

Principle of method

The polar groups of proteins bind with dyes of opposite charge causing a reduction of colour. On allowing Amido Black 10B to react with the proteins of sausage there is a decrease in the optical density at 615 nm, which, by using a standardised maceration procedure, is related to the amount of protein present[21].

Apparatus

Ultra-turrax high speed mixer, Type 18/2 Janke and Kunkel, Germany.

Reagents

Formic acid, 90%.
Buffer solution (pH 2·2) Mix 21·4 g of citric acid with 1 ml of propionic acid and dilute with water to 1 litre.
Amido Black 10B Aqueous solution standardised to an optical density of 0·70 after diluting 1 in 50 with buffer solution (see below).

Procedure

Blend 4 ± 0.01 g of comminuted sample for 1 min with about 20 ml of formic acid by means of the Ultra-turrax high speed mixer in a boiling-tube. Using formic acid, transfer the mixture quantitatively into a 50-ml volumetric flask and make the volume up to the mark with formic acid. After mixing, pipette exactly 1·0 ml of the resulting suspension into a centrifuge tube and add 0·5 ml of carbon tetrachloride. Shake the mixture vigorously for a short time to dissolve the fat. Then pipette 10 ml of dye solution into the tube and shake the mixture thoroughly for 30 s. Centrifuge the mixture for 5 min and dilute exactly 1·0 ml of the clear supernatant dye solution with buffer solution to 50 ml and measure the optical density (D_S) in a 1-cm cell at 615 nm against the buffer solution. Also, measure the optical density (D_B) of a 1 in 50 dilution of the original Amido Black 10B solution against the buffer solution. Using one brand of pork sausage:

$$\text{Protein}\ (\%)\ (\equiv N \times 6{\cdot}25) = \frac{(D_B - D_S) - 0{\cdot}122}{0{\cdot}032}$$

The above equation was found for the reagents, equipment and the particular brand of pork sausages used by the author by plotting $(D_B - D_S)$ against protein ($N \times 6{\cdot}25$) as determined by the reference macro Kjeldahl method. It is advisable, however, for each worker to derive his own equation from a similar graph produced by using his own equipment, etc. This should not present great problems in routine factory control where products made to the same recipe are examined continually.

METHOD 7.11. ESTIMATION OF ASH IN FLESH FOODS

Ash 5 g of sample in a silica dish (approximately 7 cm in diameter) at not more than 500 °C (*Method 2.7*). After weighing, retain the ash for the estimation of salt below.

METHOD 7.12. ESTIMATION OF SALT IN FLESH FOODS

Add distilled water to the ash obtained in *Method 7.11* above and mix it thoroughly with a rod. Transfer the liquid to a larger white porcelain basin down the rod and wash out the basin and rod with more water. Add 0·5–1 ml of potassium chromate solution (5 %) and titrate with 0·1 N silver nitrate solution to the first appearance of a *slight orange* colour against the yellow colour of the indicator.

$$1\ \text{ml of}\ 0{\cdot}1\ \text{N}\ AgNO_3 \equiv 0{\cdot}005845\ \text{g of NaCl}$$

METHOD 7.13. POLARIMETRIC ESTIMATION OF STARCH[22]

Reagents

Carrez solution No. 1 Zinc acetate solution (p. 60).
Carrez solution No. 2 Potassium ferrocyanide solution (p. 60).
Acidified calcium chloride solution Dissolve 620 g of calcium chloride hexahydrate crystals in 180 ml of distilled water and filter the solution until it is clear. Add a solution containing 18 g of sodium acetate trihydrate in 50 ml of water to the clear filtrate and adjust the mixture to pH 2·3 by the addition of glacial acetic acid. Adjust the specific gravity of the solution to 1·30 at 20 °C.

183

Procedure

Mix about 5 g of sample with 10 ml of water in a 400-ml tall beaker and add 50 ml of acidified calcium chloride solution. Heat the mixture in an autoclave for 10 min at a pressure of 15 lb/in^2. Cool the mixture by immersion in cold water and transfer it to a 100-ml volumetric flask by washing it with acidified calcium chloride solution until the volume is approximately 90 ml. Add 2·0 ml of Carrez solution No. 1, shake the mixture well and add 2·0 ml of Carrez solution No. 2. Shake the mixture again and dilute it to 100 ml at 20 °C with acidified calcium chloride solution. Filter the dispersion on a Whatman No. 541 filter-paper. The filtrate should be perfectly clear. Discard the first 15–20 ml of filtrate and obtain the polarimeter reading at 20 °C in a 20-cm tube on the subsequent runnings. If P is the reading at 20 °C in a 20-cm tube and $[\alpha]_D^{20}$ is 200, then calculate the amount of starch in the sample from the following expression:

$$\text{Starch } (\%) = \frac{P \times 10^4}{400 \times \text{weight taken}}$$

A gravimetric method for estimating starch has also been described[23].

The type of cereal present affects the allowance made for the nitrogen due to the filler in the calculation of the meat content (pp. 175, 176). The type of starch can be identified microscopically.

SAUSAGES

In the UK, fresh sausages are usually prepared by comminuting cuts of pork and/or beef, wheat rusk, water, salt, spices (or extracts), colour and metabisulphite preservative in a rotary bowl chopper. From time to time, various other materials, which affect the calculation of the meat content, have also been incorporated in sausage mixes, e.g., soya flour, dried milk, sodium caseinate, derived proteins, blood albumin preparations and cooked pork rinds. Polyphosphates of sodium and potassium have been used as emulsifiers as they reduce losses of fat (by rendering) during frying.

RAW MATERIALS

Meat—See p. 166.

Wheat rusk

Determine moisture ($W\%$) by drying at 100 °C, nitrogen ($N\%$) by *Method 2.6a*, fat ($F\%$) by extraction using a continuous extractor or after heating with HCl and ash ($A\%$) by igniting at 500 °C.

$$\text{Carbohydrate } (C\%) = 100 - (W + F + A + 5{\cdot}7N)$$

Calculate the N/C ratio.

Salt

Moisture Dry at 130 °C.
Assay Dissolve about 0·25 g of sample, accurately weighed, in approximately 50 ml of water and titrate with 0·1 N silver nitrate using 0·5–1 ml of potassium chromate to the first appearance of a *slight orange* colour against the yellow colour of the indicator. Calculate the percentage of NaCl in the sample.

$$1 \text{ ml of } 0{\cdot}1\text{N AgNO}_3 \equiv 0{\cdot}005845 \text{ g of NaCl}$$

Calcium and magnesium impurities Dissolve 1 g of sample in 20 ml of water, add 1 ml of dilute ammonia and 1 ml of 10% sodium phosphate solution ($Na_2HPO_4 \cdot 12H_2O$). The liquid should remain clear or give no greater turbidity than an agreed sample. Quantitative methods are described in a British Standard[24].
Sulphate Dissolve 1 g of sample in 43 ml of water, add 2 ml of dilute HCl, mix and add 5 ml of 10% $BaCl_2$ solution. Any turbidity produced after 5 min should not exceed that produced by an agreed sample.

Tests for other impurities are described in the BP. It should be borne in mind, however, that salt for food purposes would not necessarily be expected to comply with standards prescribed for pharmaceutical purposes.

Sulphite preservative

Completely dissolve 0·2 g of sample, accurately weighed, in 50 ml of 0·1 N iodine (12·69 g I + 18 g KI per litre) and then add 1 ml of concentrated HCl. Titrate the excess of iodine with 0·1 N sodium thiosulphate, adding starch solution as indicator towards the end. Calculate the result as percentage of sulphur dioxide.

$$1 \text{ ml of } 0{\cdot}1 \text{ N Na}_2\text{S}_2\text{O}_3 \equiv 0{\cdot}003203 \text{ g of SO}_2$$

It should be borne in mind that sulphites and metabisulphites slowly oxidise to sulphate during storage. Also, as SO_2 is rapidly lost in fresh sausage (approximately 100 ppm in the first day), it is advisable to use an amount in the recipe of at least 350 ppm. The legal limit is 450 ppm.

Tests for impurities are described in a BP monograph.

Herbs and spices

Determine ash, acid-insoluble ash, fixed and volatile oils, arsenic and lead (see Chapters 2–4) and examine the samples microscopically. Interpretation of results and special tests for several individual spices have been described by the author[25].

Colouring matters

Dissolve the sample in water and confirm chromatographically that only permitted colours are present (*Methods 3.4a* and *3.4b*). Also, examine the material for tinctorial power (p. 246) and lead and arsenic (see Chapter 4 and BS 3210).

Casings

Natural casings are inspected for colour, smell and freedom from knots, staining, fat and slime. Punctures can be detected by blowing them up with air or flooding them with water. Synthetic casings are produced to controlled dimensions, etc. It is important to store them in a relatively dry atmosphere.

PRODUCT

After removing any skin, prepare the sample (*Method 7.1*) and estimate the sulphur dioxide content (*Method 3.1a*, p. 79) immediately. Also, weigh out for water (*Method 7.6*) and fat (*Method 7.7*), protein (*Method 7.9/2.6a*), ash (*Method 7.11*) and salt (*Method 7.12*). Alternatively, rapid techniques including *Methods 7.8* and *7.10* are available (see also Pearson[21, 26], including the following method for sulphur dioxide).

METHOD 7.14. RAPID ESTIMATION OF SULPHUR DIOXIDE IN SAUSAGES

Principle of method

Conventional procedures for the estimation of sulphur dioxide involve distillation from acid (*Methods 3.1a* and *3.1b*). Pearson and Wong[26] have shown, however, that provided a standardised procedure is followed, the preservative can be titrated directly by applying the double titration method of Potter[27] (see *Method 3.1c*). The total reducing substances are measured by iodimetric titration of an aliquot of the filtrate obtained after macerating the sample with water, digesting with alkali and acidifying. To another aliquot, hydrogen peroxide is added prior to the titration in order to oxidise sulphite to sulphate. The difference between the titrations is equivalent to the sulphur dioxide present.

Apparatus

Mechanical macerator MSE Atomix.
Filter funnel Approximately 10 cm in diameter.
Magnetic stirrer.

Procedure

Macerate 50 g for 90 s with 250 ml of water in the MSE Atomix and filter the mixture as rapidly as possible through a large fluted filter-paper. Pipette 75 ml of filtrate into each of two flasks, A and B, followed by 75 ml of water and 5 ml of 5 N sodium hydroxide. Stir each gently with a magnetic stirrer, avoiding beating air into the solution, and allow the solutions to stand for 20 min.
Titration of A Add 7 ml of 5 N hydrochloric acid and 10 ml of 1 % starch solution and titrate immediately with 0·05 N iodine to a definite blue end-point (*A* ml).
Titration of B Add 7 ml of 5 N hydrochloric acid, 2 ml of 3 % hydrogen peroxide and 10 ml of starch solution. While stirring the mixture gently, titrate it immediately with 0·05 N iodine to a definite blue end-point (*B* ml).

$$\text{Sulphur dioxide (ppm)} = 107 (A - B)$$

The maceration time is fairly critical and is likely to vary according to the particular blender used. Also, the filtration time should be kept as short as possible. The method appears to be inapplicable to badly deteriorated samples. When any questionable result is obtained, the determination should be repeated using one of the distillation procedures.

Interpretation of results

The total meat content of sausage can be calculated from equation (7.1) on p. 176, using a factor for K_F of 0·02 (assuming wheat rusk is used) and whose N_F is 3·45 (pork)[28] or 3·55 (beef)[29]. Pork sausage should contain at least 65% of raw meat and beef sausage 50% of raw meat[30]. The lean meat content (minimum 32·5% for pork sausage and 25% for beef sausage) can be calculated from equation (7.2) on p. 177. This assumes that the meat used is a simple mixture of lean and fatty tissue, whereas in practice, of course, various cuts are used that contain different proportions of each.

Also from equation (7.1):

$$\text{Defatted meat (\%)} = DM = \frac{N_T - 0·02C}{N_F} \times 100$$

Then, the added water used in the recipe can be readily calculated:

$$\text{Added water in pork sausage (\%)} = \% \text{ water in sample} - \frac{77·4 \times DM}{100}$$

$$\text{Added water in beef sausage (\%)} = \% \text{ water in sample} - \frac{76·8 \times DM}{100}$$

Difficulties in interpretation that arise when handling samples of unknown origin due to the presence of high protein materials such as milk and soya have been discussed by the author[25]. In factory control, however, allowances can be made for such additions if necessary.

Example of calculation of meat content of pork sausage

	%
Water	49·5
Fat	27·0
Protein (N × 6·25)	10·5 ($\equiv$ 1·678% N)
Ash	2·3
Total carbohydrate (by difference)	10·7

Assuming that the nitrogen is derived from two components

only, viz., meat and wheat rusk filler, from equation (7.1) on p. 176:

$$\text{Total meat } (\%) = \left[\frac{N_T - K_F C}{N_F} \times 100 \right] + F_{Ext}$$

$$= \left[\frac{1 \cdot 678 - (0 \cdot 02 \times 10 \cdot 7)}{3 \cdot 45} \times 100 \right] + 27 \cdot 0$$

$$= \left[\frac{1 \cdot 678 - 0 \cdot 214}{3 \cdot 45} \times 100 \right] + 27 \cdot 0$$

$$= 42 \cdot 4 (\text{Defatted meat}) + 27 \cdot 0$$

$$= 69 \cdot 4$$

From the previous equations:

$$\text{Added water } (\%) = 49 \cdot 5 - \frac{77 \cdot 4 \times \% \text{ Defatted meat}}{100}$$

$$= 49 \cdot 5 - \frac{77 \cdot 4 \times 42 \cdot 4}{100}$$

$$= 17$$

From equation (7.2) on p. 177:

$$\text{Lean meat } (\%) = \frac{112 \cdot 5 \, (N_T - K_F C - B F_{Ext})}{N_F}$$

$$= \frac{112 \cdot 5 [1 \cdot 678 - (0 \cdot 02 \times 10 \cdot 7) - (0 \cdot 003833 \times 27 \cdot 0)]}{3 \cdot 45}$$

$$= 44 \cdot 5$$

Note that as the defatted meat content is greater than F_{Ext}, the calculated lean meat exceeds the defatted meat content (see p. 177).

CANNED MEATS

Meat is canned in various forms, e.g., whole, cured, types for slicing or pie fillings, or with jelly, gravy, sauce or cereal. Such products have a pH of about 6 and require relatively high processing temperatures, e.g., 240 °F. With cured meats, the curing 'reaction' may largely occur during the heat processing (p. 196). A few products, such as canned hams, are processed at considerably lower (pasteurisation) temperatures and have a shorter shelf life than other canned meats. Minimum total meat contents are prescribed for most types,

most products, a minimum lean meat content of 60% of the statutory minimum total meat content is also prescribed.

In using equation (7.1) on p. 176 for the total meat content and equation (7.2) on p. 177 for the lean meat content, it should be borne in mind that there is no correction for filler N ($K_F = 0$) if there is no cereal present or if the filler is maize or potato starch. The regulations require that the meat content is to be calculated as the total weight of meat when raw or after curing or any other similar processing as a percentage of the product.

MEAT PIES AND SAUSAGE ROLLS

Consistency in meat pie quality requires especially close control at all stages of production, including chemical, physical and micro-biological examination of raw materials and uniformity during baking, cooling, etc.[32] If 'stock' is poured through holes in the top of the pie, the gelatine used should be checked for jelly strength (p. 282) and its solution prepared by heating it sufficiently to render it commercially sterile[33], but without reducing unduly the jelly strength.

Minimum meat contents (as raw meat) are prescribed for cooked and uncooked meat pies and sausage rolls[34]. The general minima for cooked pies and sausage rolls are 25% and 12·5% of meat, respectively. As meat fat may render during cooking and become absorbed in the pastry, any fat in excess of 50% of the carbohydrate (35% in Scottish pie) in the pastry of cooked pies is considered to be part of the meat content.

For the analysis, first obtain the weights of the whole pie, the meat filling and comminuted pastry and examine them separately for water (*Method 7.6*), fat (*Method 7.7*), protein (*Method 7.9*) and ash (*Method 7.11*). Alternatively, the methods for the estimation of fat and carbohydrate in the pastry described by Dedicoat[35] can be used. The meat content of the filling (plus any allowable meat fat absorbed by the pastry in cooked products) is calculated and then expressed as a percentage of the whole pie. The necessary figures required and the basic equations for the calculation of the meat content are as follows.

	In meat filling	In pastry	In whole pie
Total weight/g	T_{Fi}	T_{Pa}	T_{Pi}
Water, %	W_{Fi}	W_{Pa}	—
Fat, %	F_{Fi}	F_{Pa}	—
Protein, %	$P_{Fi}(N \times 6{\cdot}25)$	$P_{Pa}(N \times 5{\cdot}7)$	—
Ash, %	A_{Fi}	A_{Pa}	—

	In meat filling	In pastry	In whole pie
Carbohydrate, % (by difference)	C_{Fi}	C_{Pa}	—
Meat content, %	M_{Fi}	—	M_{Pi}
Total weight of fat/g	X	Y	—
Weight of allowable meat fat/g	—	V	—

$$C_{Fi} = 100 - (W_{Fi} + F_{Fi} + P_{Fi} + A_{Fi})$$

Calculate the total meat content of meat filling (M_{Fi}) by equation (7.1) (see p. 176).

Then,

For *uncooked products:*

$$\text{Meat content } (\%) = M_{Pi} = \frac{M_{Fi}T_{Fi}}{T_{Pi}}$$

For *cooked products:*

$$\text{Weight of meat fat in pastry (g)} = V$$

$$= Y - \frac{0.5 C_{Pa}T_{Pa}}{100} \text{ (for most pies)}$$

or

$$= Y - \frac{0.35 C_{Pa}T_{Pa}}{100} \text{ (for Scottish pie)}$$

where

$$C_{Pa} = 100(W_{Pa} + F_{Pa} + P_{Pa} + A_{Pa})$$

$$\therefore V = \frac{F_{Pa}T_{Pa}}{100} - \frac{0.5 C_{Pa}T_{Pa}}{100}$$

$$= \frac{F_{Pa}T_{Pa} - 0.5 C_{Pa}T_{Pa}}{100}$$

$$\text{Meat content } (\%) = \frac{M_{Fi}T_{Fi}}{T_{Pi}} + 100 \left(\frac{F_{Pa}T_{Pa} - 0.5 C_{Pa}T_{Pa}}{100 T_{Pi}} \right)$$

$$= \frac{M_{Fi}T_{Fi} + F_{Pa}T_{Pa} - 0.5 C_{Pa}T_{Pa}}{T_{Pi}}$$

MEAT AND FISH PASTES

Pastes consist essentially of a finely comminuted mixture of meat or fish together with water, rusk, salt, seasoning and colour. The product is sealed in jars and sterilised at about 240 °F. Statutory regulations[36] prescribe that the product must contain at least 70% e.g., canned meat 95%, cured meat 90%, luncheon meat 80%[31]. For

of meat or fish (calculated as raw flesh). Also, the lean meat content of meat paste must be at least 60% of the minimum total meat content (60% of 70% = 42%). Different standards apply to potted, chopped, minced or flaked products and to brawn and dressed crab.

For the analysis, the sampling and examination for water, fat, protein, ash and salt are carried out in the same manner as that described for sausages (p. 186). Also, the total meat or fish content (as raw) can be calculated from equation (7.1) on p. 176 and, when appropriate, the lean meat content can be assessed from equation (7.2) on p. 177. Two major difficulties arise in the examination of meat pastes:

(a) Although a large number of different types of flesh are used in pastes, relatively few agreed N/FF factors have been published (at the time of writing) by the SAC. The following are available, however: beef 3·55, veal 3·35, pork 3·45, tongue 3·0, ox liver 3·45, kidney 2·7, chicken 3·7, turkey 3·65. (See also *Table 7.4*.)

(b) Elucidation of the types and proportions of the actual flesh present is difficult in a sterilised product[37].

FISH CAKES

Fish cakes are made by moulding together pre-cooked fish, mashed potato, salt and seasoning, coating with bread-crumbs and frying in batter. Statutory regulations[38] prescribe that the product must contain at least 35% of fish (calculated as raw fish). In 1951[39] and 1970[15], however, the Food Standards Committee recommended that the minimum for fish should be raised to 40%. The later Report added that the fish content of fish cakes made by large manufacturers was about 50%.

For the routine analysis, the sampling and examination for water, fat, protein, ash and salt are carried out in the same manner as that described for sausages. Assuming that white (non-fatty) fish and mashed potato are the main ingredients, the fish content can be assessed approximately from the following modified equation:

$$\text{Fish (as raw) (\%)} = \frac{\% \text{ Total N} - (0{\cdot}015 \times \% \text{ Carbohydrate})}{2{\cdot}85} \times 100$$

With samples that contain fatty fish, it is difficult to assess the proportions of the extracted fat that are due to the fish and to the frying oil, respectively. Some idea can be obtained, however, by determining the iodine value of the extracted fat (*Method 5.2*) as, in general, fish oils have a higher iodine value (e.g., salmon fat 150) than cooking oils. A more accurate allowance can be made, however, if the manufacturing recipe is known.

A new approach to the analysis of fish cakes has been presented by Burgess et al.[40], who point out that calculations based on the Stubbs and More equation are more applicable when skinned and boneless fillets are used. In practice, belly flaps, which may contain 25% of skin and bone, are frequently used.

After pressure-cooking, the softened bone can easily be crumbled and be included in the product. A further complication is that a nitrogen factor for raw fish may be inapplicable to the fish in the product as the cooking process may result in changes of composition due to loss of liquor and volatile constituents. As a result of these considerations, physicochemical methods were considered in which fish fibres are separated from other ingredients in the product before the N is determined.

METHOD 7.15. ESTIMATION OF FISH FIBRE WITH SPECIAL REFERENCE TO FISH CAKES

Reagents

Trichloracetic acid A 5% aqueous solution.
Diastase solution (1%) Dissolve the diastase in water at 35 °C.
Kjeldahl reagents See p. 52.
2.5 M Sodium acetate Containing 34·0 g of $CH_3CO_2Na·3H_2O$ per 100 ml of water.

Procedure

Weigh out 20 g of sample into a 400-ml beaker and remove the fat by stirring it three times with 25-ml portions of diethyl ether. Pour away as much of the ether as possible. Then add 80 ml of 5% trichloracetic acid, stir the mixture for 2 min and pass it through a 120-mesh sieve. Wash the sample until the washings are clear (e.g., 20 immersions each in three or four changes of water) and transfer the residue completely to a 150-ml beaker with the assistance of 75 ml of water. Heat the mixture to 85 °C, cool it to 60 °C and adjust the pH to 4·2–4·6 with 2·5 M sodium acetate solution. Then add 20 ml of the 1% diastase solution and maintain the solution at 73 °C for 30 min. Repeat with further 20-ml portions of the diastase solution until the addition of iodine solution to a small portion of the solid indicates that there is no starch remaining (no blue colour is produced). Then sieve the mixture again, wash it well, transfer it to a *weighed* 50-ml beaker, dry it at 100 °C in an oven to constant weight, then cool and weigh it. Determine the nitrogen content of the dried residue by the Kjeldahl method (*Method 2.6a,* p. 52). Use N × 6·25 for the calculation of protein.

Some of the results for the main components of fish cakes after cooking using the method are given in *Table 7.6.* Application of these and other values quoted in the paper to fish cakes did not prove to be entirely satisfactory but the authors concluded that further comparisons are necessary between their method and those based on formulae of the Stubbs and More type. Agreement is poor if much fish skin is present.

Table 7.6. RANGES OF RESULTS FOR FIBRE ESTIMATIONS ON THE COOKED COMPONENTS OF FISH CAKES

Sample	Residue, %	Nitrogen in residue, %
Cooked fish	17·1 –18·8	14·2–15·6
Cooked potato	1·4 – 3·4	2·7– 4·7
Cooked bones	10·75–12·8	10·8–13·3
Cooked skin	7·8 –10·2	13·5–14·5

MEAT CURING

Apart from salting alone, the term curing also includes the combined processes of salting and smoking. Basically, such processes cause changes in the appearance, texture and taste that make the meat more acceptable to many palates. The salt also affects the keeping qualities. Nitrite also plays a part as it combines with the myoglobin in the meat to give the familiar pink colour associated with cured meats. This and other possible changes that occur during such processes are summarised in *Figure 7.3.* The smoking process is discussed under 'Fish Curing' (p. 206).

(1) WET IMMERSION BRINING

Wet brines may contain 15–25% of salt (cf., p. 206), 0·03–0·1% of nitrite, 0·2–1·3% of nitrate, and sugars and spices. During immersion, the juices in the meat flow out and are replaced by the strong brine solution. Traditionally, nitrite may be formed by reduction of

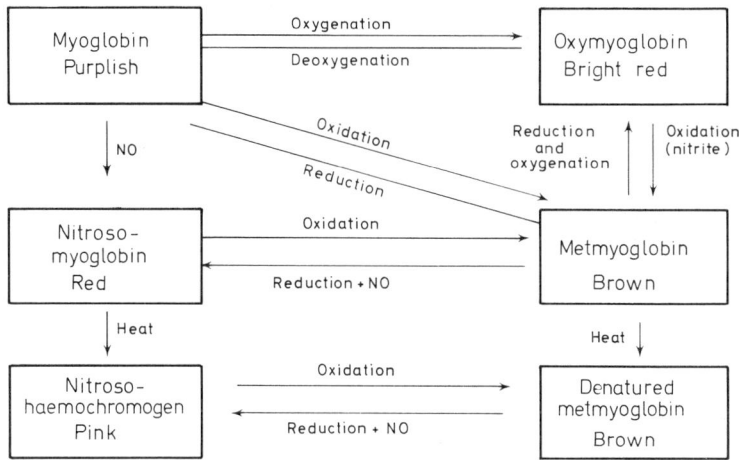

Figure 7.3. Chemical changes in the myoglobin of meat during curing, etc.

nitrate due to the presence of suitable microorganisms in the brine. A starter containing such organisms (such as a portion of an established brine) has to be added to the salt solution when such brines are freshly prepared. If nitrite is used alone, it is important to control the concentration with care, as apart from legal limitations, excessive amounts produce a fiery red colour in the product and in extreme cases may cause poisoning in the eventual consumer. The Preservative Regulations prescribe the following limits for bacon, ham, cooked and uncooked pickled meats: 200 ppm of $NaNO_2$ and 500 ppm of $NaNO_3$. The immersion time in wet curing can be reduced by applying vacuum and pressure alternately in closed vessels. A further rapid method for pre-packed bacon involves immersion of the sliced pork, then packaging it in film and finally maturation.

(2) ARTERIAL PUMPING

The arterial method involves pumping a brine containing salt and nitrite directly into the arteries. The brine is rapidly distributed through the arterial system and hence through the interior of the meat. To prevent undue saltiness in the product, the salt content of arterial brines is usually lower than that of immersion brines. The sodium nitrite concentration should preferably not exceed 500 ppm. Multiple injector needles enable the process to be shortened, and more closely controlled. The arterial method does not give efficient surface salting and it is sometimes necessary to combine the method with a subsequent immersion in wet brine.

(3) DRY CURING

Dry curing involves mixing together salt, sugar and nitrate–nitrite and rubbing the solid mixture into the meat surface. These substances dissolve in the 'weak' juices of the meat.

(4) EMULSION METHOD AND SIMILAR PROCESSES

The emulsion method, as used for the production of sausages containing cured meat, such as frankfurters, involves chopping the meat together with the salt, nitrite and ice-water. Owing to the intimate contact of the meat and curing ingredients, rapid curing is possible. A somewhat similar procedure is used for certain canned meats. For corned beef, for instance, the meat is first boiled in water (which is evaporated for the production of meat extract), mixed with the salt and nitrite, stuffed into cans and then vacuum closed and processed at 240 °F.

RAW MATERIALS USED IN MEAT CURING

SALT

Examine for assay figure of sodium chloride and for calcium and magnesium impurities as described under 'Sausages' (p. 185). It should be borne in mind that the presence of calcium and magnesium in wet brine tends to reduce the rate of salt penetration.

SALTPETRE (POTASSIUM NITRATE)

Dissolve about 0·3 g of sample, accurately weighed, in a small volume of water and transfer it to the distillation flask of a macro Kjeldahl distillation apparatus (*Figure 2.11*) using a total of 300 ml of water and add 3 g of Devarda's alloy (Cu 50, Al 45, Zn 5%). Place about 30 ml of 2% boric acid solution and screened methyl red indicator (see *Method 2.6a*) in the receiving flask and connect up the apparatus. Add 10 ml of 20% aqueous NaOH to the distillation flask through the tap funnel and distil over the ammonia formed from nitrate-N. Finally, wash down the condenser into the receiving flask with water and titrate the distillate with 0.1 N HCl or H_2SO_4 (a ml). Also, carry out a blank distillation without the sample (b ml).

$$1 \text{ ml of } 0.1 \text{ N acid} \equiv 0.01011 \text{ g of } KNO_3$$

$$KNO_3 \text{ in saltpetre } (\%) = \frac{(a-b) \times 0.01011 \times 100}{\text{Weight of sample taken}}$$

CHILI SALTPETRE (SODIUM NITRATE)

Proceed as for potassium nitrate above.

$$1 \text{ ml of } 0.1 \text{ N acid} \equiv 0.00850 \text{ g of } NaNO_3$$

SODIUM NITRITE, POTASSIUM NITRITE

Dissolve about 0.5 g of sample, accurately weighed, in a small volume of water, wash the solution into a 100-ml volumetric flask and make up the volume to the mark with water (solution A). Pipette 50 ml of exactly 0.1 N potassium permanganate into a conical flask, add 5 ml of concentrated H_2SO_4 and 100 ml of water, heat to 40 °C and titrate at about this temperature with nitrite solution A until decolorised.

$$1 \text{ ml of } 0.1 \text{ N } KMnO_4 \equiv 0.003450 \text{ g of } NaNO_2$$
$$\equiv 0.004255 \text{ g of } KNO_2$$

SUGAR, DEXTROSE—Examine polarimetrically (see p. 68).

SPICES—See p. 186.

EXAMINATION OF MEAT BRINES

SALT

The sodium chloride content of a brine can be assessed conveniently by means of a specially calibrated hydrometer (brineometer or salinometer) placed in the tank of brine. This is calibrated from $0°$ (0% NaCl) to $100°$ (26.4% w/w NaCl = a saturated solution).

$$\therefore \text{ NaCl in brine } (\% \text{ w/w}) = \frac{\text{Brineometer reading} \times 26.4}{100}$$

If other dissolved material is present, such as nitrate, the value will be higher than the actual amount of sodium chloride present. Salt is removed as more and more meat is cured in the brine, so more salt must be added to keep the proportion constant for a consistent product.

Other commercial hydrometers may be used for brines (see equivalent scales in *Figure 7.4*). A saturated sodium chloride

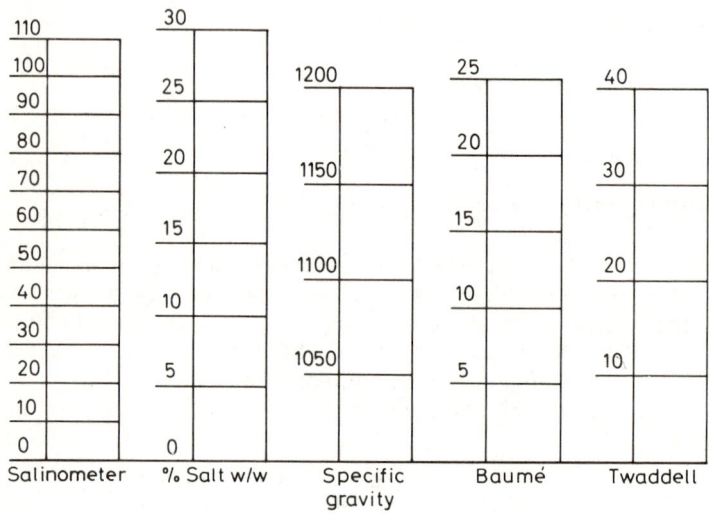

Figure 7.4. Relationships between various hydrometer scales

solution has a specific gravity of 1·204 and gives the following readings on other scales: 24·6° Baumé, 40·8° Twaddell.

In the laboratory, the sodium chloride content of the brine can be more accurately obtained by direct titration of the diluted brine with silver nitrate (using potassium chromate) or as in the following method with mercuric nitrate (using diphenylcarbazone)[41].

METHOD 7.16. ESTIMATION OF SODIUM CHLORIDE IN BRINES BY TITRATION WITH MERCURIC NITRATE

Reagents

Standard mercuric nitrate solution (approximately M/*11*) Dissolve

30 g of mercuric nitrate in 200 ml of water and 20 ml of 2 N nitric acid and dilute to 1 litre with water. Standardise the solution against sodium chloride using diphenylcarbazone as indicator by the titration method below so that 1 ml of solution $\equiv 0.01$ g of NaCl.
Diphenylcarbazone indicator 0.1 % in alcohol. Store in a refrigerator.

Method

Dilute 10 ml of brine to 500 ml with water. Pipette 25 ml of the diluted brine into a flask, add 1 ml of diphenylcarbazone indicator and titrate with the standard mercuric nitrate solution until the liquid is just violet.

NITRITE

METHOD 7.17. DETERMINATION OF NITRITE IN MEAT BRINES USING THE MODIFIED GRIESS–ILOSVAY REACTION

Principle of method

The modified Griess–Ilosvay reaction depends on the diazotisation of sulphanilic acid by nitrous acid and the coupling of the resultant compound with 1-naphthylamine-7-sulphonic acid to form a pink azo dye. The colour can be conveniently assessed by using a BDH Nessleriser.

Griess–Ilosvay reagents

Reagent No. 1 Dissolve 0.5 g of sulphanilic acid in a mixture of 30 ml of glacial acetic acid and 120 ml of water. Filter the solution.
Reagent No. 2 (modified) Dissolve 0.2 g of 1-naphthylamine-7-sulphonic acid (Cleve's acid) in a mixture of 30 ml of glacial acetic acid and 120 ml of water by warming to 50 °C. Filter the solution.
    Should either solution become coloured, shake it with a small amount of zinc dust and filter.

Apparatus

BDH Nessleriser with 50-ml Nessler glasses.
Lovibond Nessleriser nitrite disc NJ Covering the range 0.05 γ–1 γ.

Procedure

Dilute 10 ml of brine with water to give a total volume of 50 ml of solution. Pipette 2 ml of this 2% solution into a 50-ml Nessler glass, make up the volume to the mark with water and add 2 ml of Reagent No. 1 and 2 ml of Reagent No. 2. Mix and place in the right-hand compartment of the Nessleriser. Fill another Nessler glass up to the mark with distilled water and place it in the left-hand compartment. Compare the pink colour given by the sample glass with that on the disc 30 min after adding the reagents. If the colour intensity exceeds the maximum reading on the disc (1 γ), repeat the test, taking less sample. If the colour intensity is too low take a larger amount of sample. Calculate the nitrite concentration in the brine as $NaNO_2$ in ppm of the original brine.

$$1 \gamma \equiv 0.001 \text{ mg of nitrite N}$$
$$\equiv 0.00493 \text{ mg of } NaNO_2$$

Alternatively, measure the colour spectophotometrically at 520 nm. For this, assay sodium nitrite by titration against permanganate as described on p. 197 and prepare a stock solution containing 1.000 g of pure $NaNO_2$ per 100 ml of water. Prepare the working solution by diluting 5 ml of freshly prepared stock solution to 1 litre (1 ml of solution $\equiv 0.05$ mg of $NaNO_2 \equiv 50 \mu g$ of $NaNO_2$). Add by pipette to a series of 50-ml volumetric flasks 0 (as instrument blank), 0.5, 1.0, 2.0, 3.0, 4.0, 5.0, 6.0 and 7.0 ml of working solution and dilute each to the mark. Also dilute the sample solution to 50 ml. Then to each solution add 1 ml of Reagent No. 1 and 1 ml Reagent No. 2, mix and measure the optical densities, exactly 15 min after adding the reagents, in a 1-cm cell at 520 nm against the '0 ml' instrument blank. For the reference curve, plot optical density against the concentration of $NaNO_2$ in μg per 50 ml.

NITRATE

Set up two macro Kjeldahl distillation apparatus (*Figure 2.11*), one for the sample and one for the blank. Pipette 25 ml of brine sample into the distillation flask, dilute with water to 240 ml and add 3 g of Devarda's alloy. Connect up the apparatus, with 50 ml of 2% boric acid and screened methyl red indicator (see *Method 2.6a*) in the receiving flask. Add 10 ml of 20% NaOH through the tap funnel. Warm the solution *very slightly* and allow it to stand for 1 h (beware of sucking-back). Distil the solution *gently* until only 40–50 ml remains in the flask. Titrate the distillate with 0.1 N HCl or H_2SO_4.

Deduct the nitrite-nitrogen as separately determined (*Method 7.17*), allow for the blank obtained by performing the same process but omitting the sample, and calculate the final result as KNO_3 or $NaNO_3$ (% w/v). Traces of free ammonia can usually be ignored.

1 ml of 0·1 N acid $\equiv 0\cdot0014$ g of N

$$\equiv 0\cdot01011 \text{ g of } KNO_3 \text{ or } 0\cdot00850 \text{ g of } NaNO_3.$$

pH VALUE

The pH is determined directly on the undiluted brine (see p. 72). For satisfactory curing, especially if a nitrate cure is used, the pH should lie between about 5·2 and 6·2. In acid brines of pH below 5, the nitrite content is usually too low for the production of a satisfactory pink colour in the product. A pH above about 6·5 suggests that the brine is beginning to ferment and will give undesirable smelling products.

REDOX POTENTIAL

The determination of redox potential is described on p. 73. Other determinations and the interpretation of the results obtained are discussed by Hornsey and Mallows[42].

PRODUCT

SAMPLING

For comminuted and canned products and for legal purposes in general, the whole cured product is taken. In bacon curing, however, more sensitive assessments of the brine penetration can be obtained by cutting off the fat and carrying out the analyses on the minced lean tissue only. In either case, the appropriate portion should be treated as described in *Method 7.1*.

SALT

The salt can be estimated on the ash (*Method 7.12*). In routine work, however, the salt can be titrated directly on a filtered macerate.

METHOD 7.18. RAPID ESTIMATION OF SALT IN CURED MEATS

Using a mechanical macerator, blend 25 g of comminuted sample with 250 ml of water and filter. Titrate 10 ml ($\equiv 1$ g sample) with 0·1 N silver nitrate solution using potassium chromate as indicator.

$$\text{Salt}\,(\%) = \text{Titration (ml of 0·1 N AgNO}_3) \times 0·585 = S$$

The degree of salt penetration can be assessed by estimating the moisture content ($M\%$) by drying at 100 °C (*Method 7.6*) and calculating the salt as a percentage of the aqueous phase, i.e.,

$$\frac{\%\,\text{Salt}}{\%\,\text{Salt} + \%\,\text{Water}} \times 100$$

NITRITE

For some cured meats (e.g., bacon and ham, which contain no cereal), the nitrite determination can be applied directly to a dilution of 10% macerate prepared for the salt estimation above (*Method 7.18*). The modified Griess–Ilosvay reaction (*Method 7.17*) is suitable. This reagent is also applied in the following method, which involves a preliminary clearing of the solution.

METHOD 7.19. ESTIMATION OF NITRITE IN CURED MEATS

Weigh 5 g of comminuted sample into a 100-ml beaker, add 50 ml of hot water (80 °C), break up all lumps with a glass rod and wash the mixture into a 500-ml volumetric flask with 250 ml of the hot water. Place the flask on a water-bath for 2 h with occasional shaking. Then add 5 ml of saturated mercuric chloride, or 5 ml of each of the Carrez solutions (p. 60), shake the mixture, cool it, dilute it to the 500-ml mark with water and mix again. Filter and dilute a suitable portion of the filtrate to 50 ml for the test. Continue by the Nessleriser or spectrophotometric method as described in *Method 7.17*. Express the result as $NaNO_2$ in ppm.

In general, cured meat should not contain more than 200 ppm (as $NaNO_2$)[43].

NITRATE

METHOD 7.20. DETERMINATION OF NITRATE AND NITRITE IN CURED MEATS BY FOLLETT AND RATCLIFF'S METHOD[44]

Principle of method

Nitrate is determined on an extract of the meat by reduction to nitrite with cadmium and subsequent development of the dye Orange I obtained with 1-naphthol–sulphanilic acid reagent. The nitrite present as such is separately determined on the extract using the same reagent.

Preparation of reduction column

Place some zinc rods in a cold 20% solution of cadmium sulphate. After 3 h, remove the cadmium deposit and place it in a flask. Cover the material with dilute HCl and macerate it mechanically for 1–2 s and then wash it with water. Wash the resultant spongy material into the column (*Figure 7.5*). The dimensions of B are chosen to give

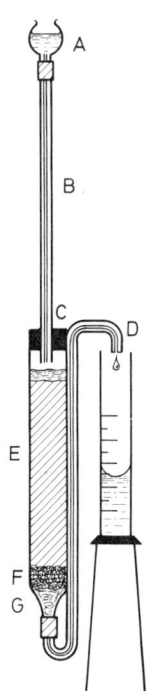

Figure 7.5. Follett and Ratcliff reduction column for the determination of nitrate: A, 25-ml thistle funnel; B, capillary tubing, O.D. 5–6 mm, I.D. 0·4 mm; C, pressure tubing; D, capillary tubing, O.D. 7–8 mm, I.D. 0·2 mm; E, reduction column, O.D. 1·4–1·5 cm, I.D. 1·2 cm; F, antibump granules; G, glass-wool plug

a flow-rate at the optimum rate of about 5 ml/min. The outlet capillary tube is designed to prevent the column running dry, so that no air is drawn into the column. Prior to a nitrate determination, wash the column with 25 ml of 0·1 N HCl followed by 50 ml of water and finally with 25 ml of buffer reagent (B) (diluted 1 + 9 with water).

Note that when the spongy cadmium is stored under water, its activity decreases after about 24 h. The reducing efficiency of the column should be regularly checked. The column can be regenerated by washing with successive portions of 0·1 N HCl, water and buffer as indicated above.

Reagents

Use nitrite-free water throughout.

A. *Potassium aluminium sulphate*—Saturated aqueous solution.

B. *Buffer solution, pH 9·6–9·7*—To 500 ml of water add 20 ml of concentrated HCl, mix, add 50 ml of ammonia of sp. gr. 0·880 and dilute the solution with water to 1 litre. Check the pH.

C. *Orange I reagent*—Warm 360 ml of water plus 50 ml of glacial acetic acid to 50 °C and pour it into a 600-ml dark bottle containing 0·25 g of powdered sulphanilic acid. Shake the mixture to dissolve the solid. Then add 0·20 g of 1-naphthol and dissolve it by shaking. Cool the solution, add 90 ml of 10% ammonia, mix and check that the pH is 4·0±0·05.

D. *Bromocresol purple*—0·05% solution in 20% alcohol.

Preparation of nitrite reference graph

Dissolve 0·50 g of $NaNO_2$ in water and dilute to exactly 1 litre. Pipette suitable volumes (0, 1, 2, 5, 10, 15, 20, 25, 30, 35 and 40 ml) into a series of 200-ml volumetric flasks. To each add 5 ml of buffer B and make up the volumes to the marks. Take 20 ml of each solution and dilute it to 100 ml with water. In each case, transfer 5 ml of solution to a 20-ml test-tube, add 10 ml of Orange I (C) and 5 ml of water and develop the colour by placing in water at 25–30 °C for 30 min. Then measure the optical density in a 1-cm cell at 474 nm against a blank consisting of 10 ml of reagent C plus 10 ml of water. Construct the reference graph by plotting optical density against concentration of nitrite (0–100 μg of $NaNO_2$) in the 5-ml solutions taken for the final colorimetric measurement. Also, express the concentrations on the graph in terms of the amounts present in the meat, assuming a 10-g sample is taken (0–2000 ppm of $NaNO_2$).

Preparation of extract of sample

Mechanically macerate 10 g of sample with 70 ml of hot water (70 °C), 5 ml of reagent B and antifoam (0·1 ml of nonanol) for 30 s. Filter the mixture through a small wire-wool plug into a 200-ml volumetric flask and wash through with more hot water. Place a thermometer in the flask, raise the temperature to 85 °C in a water-bath and then add not more than 0·5 ml of bromocresol purple (D). Add the alum reagent (A) carefully from a burette until the colour changes from purple to a definite grey (pH about 6). Cool carefully and, if necessary, add a little more of A to restore the grey colour. Then dilute to the mark with water and filter the solution through a Whatman No. 42 filter-paper. The filtrate (E), which should have a pH of 5·5–6·5, can then be used for both the determination of nitrite and nitrate.

Determination of nitrite

Pipette 20 ml of meat extract (E) into a 100-ml volumetric flask, add 5 ml of reagent B and dilute to the mark with water. Pipette 5 ml into a 20-ml test-tube, add 10 ml of reagent C and 5 ml of water and develop the colour by placing the test-tube in water at 25–30 °C for 30 min. Then measure the optical density in a 1-cm cell at 474 nm against a blank consisting of 10 ml of reagent C and 10 ml of water. Obtain the concentration of nitrite (as ppm of $NaNO_2$) in the sample by reference to the standard graph.

The amount of nitrite in bacon, ham and in cooked or uncooked pickled meat should not exceed 200 ppm as $NaNO_2$[43].

Determination of nitrate

Mix 20 ml of meat extract (E) with 5 ml of reagent B, pour the mixture into the reservoir A (*Figure 7.5*) and pass it through the column, rejecting the first 10 ml of eluate. Wash the column with water and collect 100 ml of eluate in a measuring cylinder and mix. Pipette 5 ml into a 20-ml test-tube, add 10 ml of reagent C and 5 ml of water and proceed as for the nitrite estimation. Read off the nitrate + nitrite (as $NaNO_2$) from the reference graph.

$$\text{Nitrate} = [(\text{nitrate} + \text{nitrite}) - \text{nitrite}] \times 1\cdot23$$
(as $NaNO_3$)   (as $NaNO_2$)   (as $NaNO_2$)

Allowance should also be made for the meat blank ($\equiv 13$–30 ppm of $NaNO_2$) and the reagent blank (approximately 4 ppm)—see below.

Determination of meat blank

The blank given by the meat can be assessed by removing nitrite and nitrate in a column of anion-exchange resin, Deacidite FF. The apparatus is similar to that used for nitrate reduction, except that the resin is placed in the column instead of the cadmium and the vertical capillary has a larger bore (2 mm).

Before use, pass saturated salt solution through the column, then wash it with water until the eluate is chloride-free. To 30 ml of meat extract (E), add 2 drops of concentrated HCl and pass it through the column, collecting the eluate as soon as it becomes acid, as shown by wide-range indicator paper. When the solution has passed into the column, wash it through with water. Collect 30 ml of eluate and to this add 2 drops of ammonia of sp. gr. 0·880 and shake the mixture well. Then take 20 ml of this solution and proceed as for the nitrate determination, adding 5 ml of reagent B before passing the solution through the cadmium column.

Reagent blank

Estimate the nitrite on the reagents after passing them alone through the column.

The amount of nitrate in bacon, ham and in cooked and uncooked pickled meat should not exceed 500 ppm as $NaNO_3$[43].

Other methods of determining nitrite, including the use of Griess—Ilosvay reagents, may be applied to the reduced and unreduced meat extract.

A direct method for determining nitrate specifically using brucine as colorimetric reagent has been described[45]. The AOAC determines nitrate and nitrite together after distillation, using m-xylenol as colorimetric reagent.

FISH CURING

Fish is salted by either the wet or dry method. Only salt is normally used, i.e., unlike meat, nitrite and nitrate are not included. Saturated brine is often used. Also, for some smoked fish, coal-tar dyes such as Brown FK (for kippers) and amaranth or tartrazine (for white fish) are incorporated in wet brine in order to attain the 'smoked' colour in the product with a shorter smoking time.

As uniformity is difficult to attain in traditional chimney-type kilns, mechanical kilns have been designed to give control of the ventilation and the temperature density and rate of flow of the smoke[46, 47]. Owing to the more general use of shorter smoking times, smoked fish nowadays is not dried to the same extent as it was formerly. The process is therefore not now considered to be so much a method of preservation. It does, however, cause several antiseptic substances to be deposited on the fish surface, such as aldehydes, ketones, alcohols, acids, phenols and tars. Most products, including kippers, bloaters and finnan haddock, are cold-smoked at less than 100 °F and contain 2–3% of salt. Buckling from herring and smokies from haddock are, however, hot-smoked at over 200 °F so that the centre temperature reaches over 140 °F and the fish is cooked.

A more recent development is the use of smoke dips. This involves drying the fish after dipping it in an aqueous extract of smoke constituents.

For routine work, cured fish should be examined for water (*Method 7.6*) and salt (*Method 7.18*).

MERCURY IN CANNED TUNA

In December 1970 the US Government announced that more than one can of tuna in five tested contained a level of mercury in excess of the limit of 0·5 ppm. Immediately afterwards in the UK the Government Chemist announced that 21·6% of the samples of canned tuna examined contained mercury in excess of the US limit. Then, from a survey in 1971, the APA reported a rather lower figure of 5·2%. This contamination is probably acquired by fish which live in water into which factory waste containing mercury is discharged. In the previous 20 years, more than 100 persons were reported to have died in Japan due to eating fish containing about 20 ppm of mercury derived from discharge from chlorine plants. In such fish the mercury is metabolised in the flesh to the cumulative and toxic form of methyl mercury.

In the analysis for mercury it is usual for routine purposes to estimate the *total* mercury by colorimetric or atomic absorption techniques. The following general method using dithizone is the reference colorimetric procedure recommended by the SAC[48].

METHOD 7.21. DETERMINATION OF MERCURY IN FOODS SUCH AS CANNED TUNA USING DITHIZONE[48]

Principle of method

The organic matter in the sample is destroyed by wet oxidation taking special precautions to trap volatile mercury. The acid solution is reduced with hydroxylammonium chloride to destroy oxides of nitrogen and the mercury is extracted with dithizone. The mercury is returned to the aqueous phase by oxidation with acid nitrite, the extracts. In the presence of over 60 μg of copper, chloroform is used any remaining oxides of nitrogen are removed with urea. After adding EDTA to hinder the reaction of copper with dithizone, the mercury is extracted stepwise with dithizone in carbon tetrachloride. The mercury is then determined at 485 nm on the diluted combined extracts. In the presence of over 60 μg of copper, chloroform is used instead of carbon tetrachloride and the measurement made at 492 nm.

Apparatus

Digestion apparatus—See Figure 7.6.

Reagents

Hydroxylammonium chloride Shake a 20% w/v aqueous solution with a few ml of dithizone stock solution in a separator. After allowing the layers to separate, reject the organic layer. Repeat the extraction until the organic layer has only the colour of pure dithizone solution. Finally, extract the aqueous solution with successive small amounts of chloroform until the extracts are colourless and discard the extracts.
Dithizone stock solution Prepare a 0·05% w/v solution in chloroform.
Dilute dithizone solution in chloroform Dilute 2 ml of dithizone stock solution to 100 ml with chloroform.
Dilute dithizone solution in carbon tetrachloride Dilute 2 ml of dithizone stock solution to 100 ml with carbon tetrachloride.
*Hydrochloric acid, 0.1 N
*Sodium nitrite Prepare a 5% w/v aqueous solution.
*Urea Prepare a 10% w/v aqueous solution
*EDTA Dissolve 2·5 g of EDTA (disodium salt dihydrate) in 100 ml of water

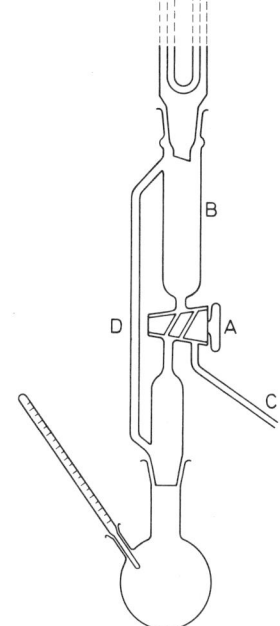

Figure 7.6. Special all-glass apparatus for the wet oxidation of materials where the metal is liable to volatilise during the heating: The flask has a capacity of 250 ml and the reservoir B of 150–200 ml. The reservoir is fitted with a reflux condenser. The thermometer should read up to 200 °C. Ground-glass joints must be used throughout.

*Acetic acid Approximately 4 N.

Standard mercury stock solution Dissolve 0·1354 g of mercuric chloride in 1 litre of 0·1 N HCl.
Dilute standard mercury solution Dilute 10 ml of mercury stock solution to 1 litre with 0·1 N HCl.

$$1 \text{ ml of solution} \equiv 1 \ \mu g \text{ of Hg}$$

Procedure

(*N.B.* Perform a blank at the same time.) To a suitable quantity of sample (e.g., 5 or 10 g canned fish) in the oxidation flask (*Figure 7.6*) add a cold mixture of water (up to 20 ml according to the water content of the sample), 5 ml concentrated sulphuric acid and 50 ml concentrated nitric acid (if the dry matter in the sample exceeds 10 g add a further 5 ml nitric acid for each gram in excess). Add glass

* Purify if necessary by shaking with dithizone as described for hydroxylammonium chloride.

beads and assemble the apparatus. Close tap A and heat the flask, gently at first, and collect the distillate in B. When the temperature on the thermometer reaches 116 °C open A and collect the distillate in a measuring cylinder from the drain tube C. Continue collecting the distillate in B and, when the oxidation mixture darkens, run a little of the distillate from B into the flask. Continue in this way, maintaining a slight excess of nitric acid in the flask, until the liquid no longer darkens and fumes of sulphuric acid are evolved. Then allow to cool, run the liquid in B into the flask and add to the first distillate in the measuring cylinder (volume of residue plus distillate is usually just under 100 ml). After titrating 1 ml of the solution with standard NaOH to determine its normality, dilute the whole solution with water to an approximate acidity of N. (Total volume should then be about 400 ml.) Then heat to boiling, remove the heat source and add, rapidly with mixing, an amount of hydroxylammonium chloride solution equal to 0·1 of the total bulk and cool.

Transfer the solution to a 1-litre separating funnel and remove any fat by shaking with carbon tetrachloride. Then add 10 ml of the dilute dithizone in carbon tetrachloride, shake for 1 min, allow to separate and run the lower layer into a 150-ml separator. Continue extracting with 1-ml quantities of the dithizone solution until two successive extracts remain green. Combine the extracts in the 150-ml separator, add 10 ml of 0·1 N HCl and 1 ml sodium nitrite solution, shake vigorously for 1 min and, after separating, discard the lower layer. Add 1 ml of hydroxylammonium chloride solution, set aside with occasional shaking for 15 min, and then add 1 ml urea solution and 1 ml of EDTA solution.

Add 0·5 ml of dilute dithizone in carbon tetrachloride from a 10-ml burette. If copper is present, a solution of dithizone in chloroform should be used and the final readings should then be taken at a wavelength of 492 nm. Shake the separator vigorously for 10 s, run the separated lower layer into a further separator containing 5 ml of 4 N acetic acid and repeat the operation until the separated layer is greenish orange. Then extend the shaking time to 30 s and reduce the additions of dithizone solution to 0·2 ml. Continue the titration and separation, combining the extracts, until the organic layer has a greyish mixed colour, indicating that the mercury has been extracted completely and also that the extract contains a slight excess of dithizone. Note the volume of dithizone required. From another 10-ml burette add sufficient carbon tetra-chloride or chloroform to adjust the volume of extract to 4·0 ml. Then mix, dry the stem of the funnel and run the lower layer through a small glass-wool plug (supported in a small glass funnel) into a 1-cm cell and measure the optical density at 485 nm against the blank in the reference cell. Compare the reading with the calibration graph to obtain the concentration of mercury in the sample.

Preparation of calibration graph

To a series of separators add aliquots of the dilute standard mercury solution to cover the range 0·0–10·0 μg of mercury and dilute each to 10·0 ml with 0·1 N HCl. Treat each in the following manner: add 1 ml of sodium nitrite solution and 1 ml of hydroxylammonium chloride solution, mix, set aside for 15 min and add 1 ml of urea solution and 1 ml of EDTA solution. Then complete the extraction and measurement of each as for the sample using the same batch of dithizone. Construct the calibration curve relating optical density at 485 nm to the number of μg of Hg in each separator.

A rapid method for determining total mercury using flameless atomic absorption spectrophotometry has been described by Uthe, Armstrong and Stainton[49]. Methods for the more specific determination of organic mercury compounds including methyl mercury using TLC–GLC techniques have been described by Westoo[50, 51] and Tatton and Wagstaffe[52].

REFERENCES

1. THORNTON, H., *Textbook of Meat Inspection*, 5th edn, Bailliere, Tindall & Cassell, London (1968)
2. GERRARD, F., *Meat Technology*, 5th edn, Leonard Hill, London (1969)
3. *The Distinguishing Features of Fish*, The Worshipful Company of Fishmongers, London (1958)
4. PEARSON, D., *J. Sci. Fd Agric.*, **19**, 366 (1968)
5. LUCKE, F. and GEIDEL, W., *Z. Unters. Lebensmittel*, **70**, 441 (1935)
6. PEARSON, D. and MUSLEMUDDIN, M., *J. Assoc. Publ. Analysts*, **6**, 117 (1968)
7. PEARSON, D. and MUSLEMUDDIN, M., *J. Assoc. Publ. Analysts*, **7**, 50 (1969)
8. PEARSON. D. and MUSLEMUDDIN, M., *J. Assoc. Publ. Analysts*, **7**, 73 (1969)
9. PEARSON, D. and MUSLEMUDDIN, M., *J. Assoc. Publ. Analysts*, **9**, 28 (1971)
10. FERNANDEZ-FLORES, E. and SALWIN, H., *J. Ass. off. analyt. Chem.*, **51**, 1109 (1968)
11. PEARSON, D., *J. Sci. Fd Agric.*, **19**, 553 (1968)
12. CHARNLEY, F. and DAVIES, F. R. E., *Analyst, Lond.*, **69**, 302 (1944)
13. TARLADGIS, B. G., WATTS, B. M., YOUNATHAN, M. T. and DUGAN, L., *J. Am. Oil Chem. Soc.*, **37**, 44 (1960)
14. SINNHUBER, R. O., YU, T. C. and YU, Te C., *Fd Res.*, **23**, 626 (1958)
15. *Food Standards Committee Report on the Pre-1955 Compositional Orders*, H.M.S.O., London (1970)
16. MAHMOOD-UL-HASSAN and PEARSON, D., *J. Sci. Fd Agric.*, **17**, 421 (1966)
17. SMITH, P. R., 'A Review of Rapid Methods for the Estimation of Total Fat', *BFMIRA Scientific and Technical Surveys*, No. 56 (1969)
18. BEN-GERA, I. and NORRIS, K. H., *J. Fd Sci.*, **33**, No. 1, 64 (1968)

19. DAVIS, C. E., OCKERMAN, H. W. and CAHILL, V. R., *Fd Technol., Champaign*, **20**, 1475 (1966)
20. WHITEHEAD, R. C., *Fd Technol., Champaign*, **24**, 165 (1970)
21. PEARSON, D., *Fd Mf.*, **47**, No. 4, 45 (1972)
22. FRASER, J. R. and HOLMES, D. C., *Analyst, Lond.*, **83**, 371 (1958)
23. SOCIETY OF PUBLIC ANALYSTS, *Analyst, Lond.*, **77**, 544 (1952)
24. *Specification for Vacuum Salt for Butter and Cheese Making and Other Food Uses*, BS 998:1969, British Standards Institution, London
25. PEARSON, D., *The Chemical Analysis of Foods*, 6th edn, Churchill, London (1970)
26. PEARSON, D. and WONG, T. S. A., *J. Fd Technol.*, **6**, No. 2, 179 (1971)
27. POTTER, E. F., *Fd Technol., Champaign*, **8**, 269 (1954)
28. SOCIETY FOR ANALYTICAL CHEMISTRY, *Analyst, Lond.*, **86**, 557 (1961)
29. SOCIETY FOR ANALYTICAL CHEMISTRY, *Analyst, Lond.*, **88**, 422 (1963)
30. *The Sausage and Other Meat Product Regulations 1967 (SI 1967 No. 862; 1968 No. 2047)*, H.M.S.O., London
31. *The Canned Meat Product Regulations 1967 (SI 1967 No. 861; 1968 No. 2046)*, H.M.S.O., London
32. WILLIAMS, E. F., 'Meat and Meat Products', in HERSCHDOERFER, S.M., (Editor), *Quality Control in the Food Industry*, Vol. 2, Academic Press, London and New York, 251 (1968)
33. MINISTRY OF FOOD, *Report of the Manufactured Meat Products Working Party*, H.M.S.O., London, 27 (1950)
34. *The Meat Pie and Sausage Roll Regulations 1967 (SI 1967 No. 860)*, H.M.S.O., London
35. DEDICOAT, H., *J. Assoc. Publ. Analysts*, **1**, No. 4, 85 (1963)
36. *The Fish and Meat Spreadable Products Regulations 1968 (SI 1968 No. 430)*, H.M.S.O., London
37. MACKIE, I. M. and TAYLOR, T., *Analyst, Lond.*, **97**, 609 (1972)
38. *The Food Standards (Fish Cakes) Order 1950 (SI 1950 No. 589)*, H.M.S.O., London.
39. *Food Standards Committee Report on Fish Cakes*, H.M.S.O., London (1951)
40. BURGESS, G. H. O., MCLACHLAN, T., TATTERSON, I. N. and WINDSOR, M. L., *Analyst, Lond.*, **95**, 471 (1970)
41. SCHONHERZ, Z., *Fd Mf.*, **30**, 460 (1955)
42. HORNSEY, H. C. and MALLOWS, J. H., *J. Sci. Fd Agric.*, **5**, 573 (1954); **6**, 705 (1955)
43. *The Preservatives in Food Regulations 1962 (SI 1962 No. 1532) (Also 1971 Amendment, SI 1971 No. 882)*, H.M.S.O., London
44. FOLLETT, M. J. and RATCLIFF, P. W., *J. Sci. Fd Agric.*, **14**, 138 (1963)
45. LANDMANN, W. A., SAEED, M., PIH, K. and DOTY, D. M., *J. Ass. off. agric. Chem.*, **43**, 531 (1960)
46. CUTTING, C. L. and SPENCER, R. J., 'Fish and Fish Products', in HERSCHDOERFER, S. M., (Editor), *Quality Control in the Food Industry*, Vol. 2, Academic Press, London and New York, 303 (1968)
47. ZAITSEV, V., KIZEVETTER, I., LAGUNOV, L., MAKAROVA, T., MINDER, L. and PODSE-VALOV, V. (Translated by A. de Merindol), *Fish Curing and Processing*, MIR Publishers, Moscow (1969)
48. SOCIETY FOR ANALYTICAL CHEMISTRY, *Analyst, Lond.*, **90**, 515 (1965)
49. UTHE, J. F., ARMSTRONG, F. A. J. and STAINTON, M. P., *J. Fish. Res. Bd Can.*, **27**, 805 (1970)
50. WESTOO, G., *Acta chem. scand.*, **20**, 2131 (1966)
51. WESTOO, G., *Acta chem. scand.*, **21**, 1790 (1967)
52. TATTON, J. O. and WAGSTAFFE, P. J., *J. Chromatog.*, **44**, 284 (1969)

8

FLOUR, BAKING POWDER AND
SELF-RAISING FLOUR

FLOUR

Wholemeal (brown) flour is made essentially by grinding the whole
wheat grain and therefore includes a mixture of outer branny skins
(13%), the embryo (2%) and the inner starchy endosperm (85%)
(*Figure 8.1*). During milling, the grains are split, the endosperm is

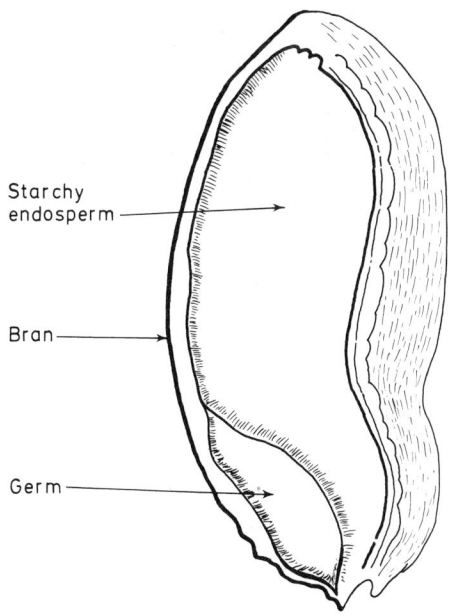

Figure 8.1. Structure of the wheat grain

separated from the skins and embryos and a flour of increasing whiteness is produced. A typical white flour has a 72% extraction rate, i.e., 72 parts of flour are produced from 100 parts of wheat.

Before milling, the cleaned wheat is conditioned to a moisture content (using the Carter–Simon method of drying at 155 °C—*Method 2.2d*, p. 33) of about 15% (for soft European) to 19% (for hard Canadian). To achieve this, home wheats may have to be dried and imported varieties 'hot conditioned', involving adding water and heating the moist wheat to about 50 °C in order to encourage distribution of the water throughout the kernels. In general, conditioning facilitates sieving so that the skins of the kernels are tough and not brittle and the endosperm friable.

In the milling process, the grains are passed to the break rolls, which consists of pairs of corrugated rollers revolving at different speeds. The extracted endosperm (semolina) is sieved away and the split grains are passed to other break rolls that open them further and more endosperm is scraped away. The break-roll treatment and sieving are repeated several times, together with aspiration, and purified flour is eventually obtained.

Bran contains more ash and fibre than the starchy endosperm, so these contents can be used to assess the extraction rate of the flour. Similarly, owing to the greater amount in the bran, brown flour contains more iron and vitamins than white flour, so the latter has to be fortified with reduced iron, vitamin B_1 and nicotinic acid to ensure that it complies with the relevant legal requirements[1]. Regulations also prescribe, except in the case of wholemeal, the addition of chalk to flour. Such fortification, even when the nutrients are introduced as a 'master mix', causes problems in attaining homogeneity throughout the batch and hence from packet to packet. The reasons for fortifying flour and the problems associated with it have been discussed by the FSC[2].

EXAMINATION OF FLOUR

The examination of flour varies considerably. Apart from the different requirements as to the use to which it is to be put, flour is a product in the mill, but a raw material in the bakery. Apart from legal requirements, flour is used for different products and this may affect the choice of tests as well as their interpretation. Publications that give details of the specialised tests that can be applied to flour are available[3,4]. In the main, the procedures discussed in this Chapter can be performed in the average laboratory, i.e., few of the methods described require specialised equipment.

MOISTURE

The moisture content of flour varies according to the conditioning of the wheat as well as the climatic conditions. From Section 2.2 (p. 31), it is apparent that wheat flour contains much bound water and the value obtained from drying methods varies considerably according to the drying temperature used. Conditions that are commonly used include drying at 100 °C for 5 h (*Method 2.2b*), at 130 °C for 1 h or 155 °C for 15 min (Carter–Simon method, *Method 2.2d*). Using the latter method, the moisture content of flour is normally between 12 and 15%. Electrical meters (p. 36) are also most useful for obtaining rapid results at the mill, during receipt of raw materials and for warehouse control. Only flours with a relatively low moisture content should be used for the preparation of cake mixes and self-raising flour.

ASH

The ash of flour is obtained by incineration at 550 °C (*Method 2.7*). As the wheat skins have a much higher mineral content than the inner endosperm, the ash content gives some idea of the flour grade. Thus, whereas brown wholemeal flour gives an ash content of about 2%, unfortified white (72% extraction) flour gives an ash content of about 0·4%. Some cake flours give an ash content as low as 0·3%. In view, however, of the legal requirement[1] that all flour (except wholemeal) must be fortified with 235–390 mg of $CaCO_3$ per 100 g, it is more difficult than formerly to assess the grade from the ash figure (see also p. 222).

FIBRE

The fibre of flour can be estimated by *Method 2.5* (including Notes *f*, *j* and *l*, p. 49) or by the AOAC method. As the wheat skins have a much higher fibre content than the endosperm, this value also gives an indication of the flour grade. Wholemeal flour has a fibre content of about 2%, light wholemeals contain approximately 0·6%, and 72% extraction flours may contain as little as 0·1–0·2%. In order to ensure that 'wheatmeal' and 'brown' flours contain a reasonable proportion of 'skin', they are required by law[1] to contain at least 0·6% of fibre (% w/w of the dry matter). Although useful for assessing the flour grade, especially as the value obtained is (unlike the ash) unaffected by fortification with chalk, the estimation

215

of fibre possesses the practical disadvantage that the estimation is somewhat cumbersome to apply continually in routine work (cf., van Soest[5]).

PROTEIN

Protein in flour is estimated by the Kjeldahl procedure using 1·5 g of sample for the macro method, *Method 2.6a*, 0·15–0·2 g for the semi-micro method, *Method 2.6b(a)*, or 2 g for *Method 2.6b(b)*. The protein is calculated from the factor $N \times 5·7$ (cf., Tkachuk[6,7]). Strong bread flours should contain 11–12% of protein, but those suitable for cake making should be weak flours containing less than 10% of protein. Protein in flour can also be estimated rapidly by using absorptiometric methods[8–10].

GLUTEN

The strength of a flour depends to a considerable extent on the nature and amount of gluten present. Such properties can be assessed by forming a dough with water, washing out the starch and examining the elastic mass remaining. Two methods are described below —one involving hand washing, the other machine washing. The crude gluten obtained is not pure protein, but contains starch, lipids and mineral matter. Difficulties associated with the interpretation of the test are discussed by Kent-Jones and Amos[3].

METHOD 8.1A. EXAMINATION OF CRUDE GLUTEN IN FLOUR (HAND-WASHING METHOD)

Place 25 g of flour in a mortar, add about 15 ml of tap water, work the mixture into a dough with a pestle or spatula and allow it to stand for 1 h. Knead the dough gently under a water tap (and over a No. 60 sieve) for 10–15 min so that soluble matter and starch are washed away. Place the solid matter under water for 1 h and then compress it as dry as possible, roll it into a ball, weigh the moist gluten and express it as a percentage of the flour. Also note its colour, toughness and elasticity. Then dry it at 100 °C to constant weight, weigh it and express it again as a percentage of the flour.

METHOD 8.1B. EXAMINATION OF CRUDE GLUTEN (SIMON AUTOMATIC GLUTEN WASHER)

The machine has a kneading chamber with serrated walls and a perforated base covered with No. 50–60 silk. Water is fed into the chamber through a water column, which can be supplied either by gravity or by a siphon pipe from a container calibrated in divisions of 500 ml. The kneading chamber contains a rotor with four beryllium–copper spring arms, the rotor being driven through reduction gears by a constant-speed electric motor. A time-clock is fitted to control the duration of the washing period and to switch off the motor on its completion.

Procedure

Thoroughly moisten the silk at the base of the kneading chamber and adjust the water supply. Mix 10 g of flour with 5 ml of water to give a clear dough by rolling for 1 min on a smooth surface (preferably a glass plate) to form a smooth ball. Place the dough between two of the rotor arms in the kneading chamber and turn the time-clock to the required washing time (usually 15 min). After the washing, remove the gluten, rinse it under the tap, dry it with a sponge, obtain the wet gluten weight and note the colour, toughness and elasticity. Dry the wet gluten by heating it in a weighed moisture dish in the Carter–Simon oven (*Method 2.2d*) and weigh it again. Calculate the percentages of dry and wet gluten in the flour.

OIL

As wheat skins contain much more oil than the endosperm, wholemeal flour contains more (about 2%) oil than does white 72% extraction flour (about 1%). Direct treatment of the sample with ether or light petroleum extracts the free oil only, so it is preferable to use the following acid digestion procedure, which determines the free plus combined oil.

METHOD 8.2. DETERMINATION OF OIL IN FLOUR

Weigh out accurately about 2 g of flour into a 100-ml beaker, add 2 ml of alcohol and stir the mixture with a small rod. (To facilitate dispersion of the sample, stir it first with the minimum volume of water and a few drops of ammonia of sp. gr. 0·880. Then add 2 ml of alcohol. Keep the water–concentrated HCl ratio as near to 3:7 as possible before heating to 80 °C.) Add 3 ml water and 7 ml of concentrated HCl and stir the mixture. Place it in water at 80 °C (stirring frequently) for about 30 min in order to complete the hydrolysis. Cool the mixture and transfer it (down the rod) into a separating funnel with the assistance of 10 ml of alcohol and then 25 ml of diethyl ether. Stopper the separator, shake it, add 25 ml of light petroleum (boiling range 40–60 °C) and shake it again. Then continue as in *Method 6.20* commencing with 'After the layers have separated, . . .'.

ACIDITY AND pH

The acidity of flour increases during storage, particularly if it is stored at too high a temperature, with consequent impairment of baking quality. Also, the acidity of brown flour is higher than that of white flour prepared from the same grain.

Methods for estimating the acidity on a water extract (direct or defatted) or an alcoholic extract are being replaced by titration of the free fatty acids of the extracted fat. The value of the latter increases more rapidly during storage than values of properties determined in other methods.

METHOD 8.3. ESTIMATION OF FAT ACIDITY IN FLOUR

Solvent mixture—Mix 50 ml of benzene and 50 ml of alcohol, add 1 ml of 1% phenolphthalein in alcohol and add, drop by drop, 0·05 N aqueous NaOH until the solution is a faint pink colour.

Procedure

Extract 10 g of finely ground sample with light petroleum (boiling range 40–60 °C) in a Bolton extractor (*Figure 2.7*) for 4 h. Remove the solvent (*Figure 6.3*), add 50 ml of solvent mixture and titrate with 0·05 N NaOH until the solution is a faint pink colour. Express the fat acidity as millilitres of 1 N alkali required per 100 g of flour. Alternatively, from the weight of fat titrated, calculate the FFA as a percentage of oleic acid on the extracted fat (p. 125).

The interpretation of the results obtained has been discussed by Morrison[11, 12].

The pH (p. 72) is best measured by macerating 10 g of flour with 100 ml of water, allowing the mixture to stand for 30 min and decanting off the supernatant liquid for the determination. The method is of most value in the control of treated cake flours, as chlorine treatment normally causes a drop in pH from about 6·5 to below 5.

PEKAR COLOUR TEST

At the mill, the grade of a flour can be readily assessed by visual comparison against that of flour of known extraction rate. In the Pekar test, the flour is compressed between glass plates. As the tint obtained varies according to factors such as the thickness of the flour layer, the pressure applied and the moisture content, it is necessary to apply the same technique to both the test and the comparison sample[3].

METHOD 8.4. ASSESSMENT OF FLOUR COLOUR AND GRADE USING THE PEKAR COMPARISON TEST

Apparatus—Rectangular glass plates, 12 × 8 cm and about 2–3 mm thick.

Procedure

Weigh out about 12 g of flour and pack it on one side of one of the glass plates in a straight line with the assistance of another plate. Treat 12 g of the standard flour of known comparable extraction rate in the same manner so that the straight edges of the two flours are adjacent. Carefully move one of the portions so that it will be in contact with the other and slick both with one stroke of the spatula so that the thickness of the layer diminishes from about 0·5 cm in the middle of the plate to a thin film at the edge. The line of demarcation between the two flours should be quite distinct. Note any differences in colour between the two samples and repeat if necessary with flours of different and more comparable extraction rates. Also, dip the slabs in cold water and make further comparisons immediately after dipping and 1 h later. A further useful comparison is possible by pouring a 0·2 % alcoholic solution of catechol on to the slab after dipping.

BRIGHTNESS AND ASSESSMENT OF FLOUR GRADE

The difficulties associated with the assessment of the extraction rate from the ash and fibre values have been referred to previously. The grade of flour can be readily assessed by means of the Kent-Jones and Martin Flour Colour Grader[13]. This measures the diffuse reflection of light at 530 nm of a paste made from the flour. At this wavelength, the reading obtained is a measure of the bran content and is almost unaffected by bleaching, which affects the yellow xanthophylls in the fat. Also, the use of a paste nullifies the apparent differences in visual greyness due to variations in particle size. The commercial instrument on which the following method is based is manufactured by Henry Simon Ltd., Stockport, England. In the U.S.A., the 'Agtron', based on similar principles, is used for assessing the colour and grade of flour[14].

METHOD 8.5. THE ASSESSMENT OF FLOUR GRADE USING THE
KENT-JONES AND MARTIN FLOUR COLOUR GRADER (SERIES 2)

The position of the main controls on the top of the instrument are shown in *Figure 8.2.*

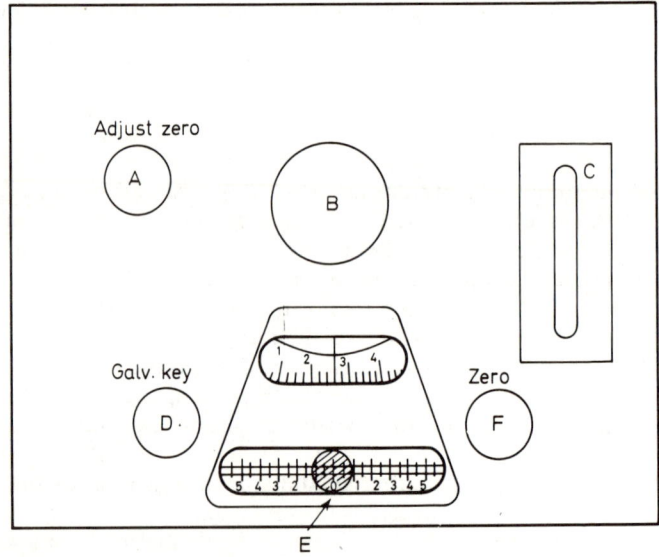

Figure 8.2. Position of controls on the Kent-Jones and Martin Flour Colour Grader (Series 2)

Procedure

Switch on the instrument and allow it to warm up for 30 min. Open the door on the front of the instrument and withdraw the magnet–galvanometer assembly to give access to the clamping screw on the centre of the rear side of the assembly. Unscrew this to release the galvanometer, replace the assembly and close the door. Bring the light spot, E, approximately to the centre position by careful movement of the black milled knob on the top of the galvanometer assembly. Adjust to exact zero by turning the Zero knob, F. Raise the cell holder, C, gently to the upper position (about 4 in) by means of the handle. Rotate knob B until the red line on the dial and the black lines on the cursor coincide. Press the Galv. Key, D, and continue tapping it whilst rotating the Adjust Zero control, A, until there is no movement of the light spot away from the zero. Bring the spot exactly to zero by using the Zero control, F. Obtain a cell and cap, which are stored in the right-hand side of the instrument, and close the right-hand door after removing them. Prepare the sample by mixing 30 g of flour with 50 ml of water to produce a smooth paste. *It is recommended that the reading should be taken 90 s after starting the mixing.* Pour the paste into the cell, holding the latter at an angle so that the frosted side is uppermost (to keep air bubbles away from the testing surface). Fill the cell to between the two lines. Close the cell with the cap. Insert the cell (frosted side outwards) in the upper part of the holder and clamp it by pressing the clamping bar downwards. Check that the galvanometer light spot reads zero on tapping D—otherwise restore to zero as above. Push down the cell holder C into its lower position, tap the Galv. Key, D, and rotate B until the light spot again returns to zero. Read off the grade figure as follows:

Type of flour	Grade figure
Patent (high grade)	−1·0 to 1·5
72 % Extraction	2·0 to 4·5
80 % Extraction	5·0 to 7·5
85 % Extraction	8·0 to 12·5

Raise the cell holder, remove the cell and clean it. Replace the cell in the right-hand side of the instrument.

Self-raising flours should be prepared by mixing 30 g of sample with 50 ml of buffer solution prepared as follows.

Dissolve 110 g of anhydrous disodium hydrogen phosphate and 77 g of citric acid in 1 litre of distilled water. Dilute 20 ml of solution to 100 ml with water just prior to the test.

The Grade Figure can also be calculated approximately from the ash of unfortified flour:

Flour Grade Colour Figure (Kent-Jones and Martin) =
$$(14\cdot8 \times \% \text{ Ash}) - 4\cdot8$$

If the amount of chalk present in a fortified flour is subtracted from the total ash, the above equation can still be used. When the chalk content is unknown, a subtraction of $0\cdot3\%$ from the total ash can be used.

BLEACH VALUE

The bleach value is usually used as a measure of the extent by which the creaminess of flour is reduced by the application of bleaching agents. After extraction with a benzene–petrol mixture, the yellowness of the unoxidised xanthophylls can be measured spectrophotometrically.

METHOD 8.6. DETERMINATION OF THE BLEACH VALUE OF FLOUR

Extraction reagent—Mix 100 ml of petrol with 100 ml of benzene and add a few drops of alcohol to remove any turbidity.

Procedure

Shake 18 g of flour frequently with 100 ml of extraction reagent in a stoppered reagent bottle over a period of 3 h. Filter the mixture rapidly, rejecting the first runnings. Measure the optical density of the clear filtrate in a 4-cm cell at 425 nm (D_{425}) and 660 nm (D_{660}) using the extraction reagent for setting the instrument. Calculate $D = D_{425} - D_{660}$. A heavily bleached flour gives a low value for D, below $0\cdot1$, whereas unbleached flour gives a value above $0\cdot2$.

In the U.S.A., the colour is extracted with n-butanol, the optical density is measured at $435\cdot8$ nm and the result (for 'Pigments in Flour') is expressed in terms of carotene.

PARTICLE SIZE

The granularity of flour can be assessed by placing 100 g of flour in the top compartment of a nest of sieves of various mesh sizes. Each sieve contains two rubber balls and, after closing, the whole nest of sieves is shaken (preferably mechanically) for 5 min. The distribution of particle sizes is then obtained by expressing the weight remaining on each sieve as a percentage of the 100 g of sample taken. The particle size can also be assessed from the sedimentation rate[4, 15, 16].

DIASTATIC ACTIVITY

When a bread flour is mixed with water, salt and yeast, the dough produced should provide the yeast with sufficient sugar for adequate production of gas and should expand adequately owing to the pressure of the gas. A breadmaking flour should therefore contain an adequate amount of protein, have suitable physical properties and be able to convert starch into sugar as measured by the diastatic activity. The principal diastatic enzymes present are α- and β-amylase. The method described below is essentially that due to Blish and Sandstedt[17].

METHOD 8.7. ESTIMATION OF THE MALTOSE VALUE OF FLOUR USING BLISH AND SANDSTEDT'S METHOD[17]

Reagents

Alkaline ferricyanide solution (0·1 N) Dissolve 33 g of potassium ferricyanide and 44 g of anhydrous sodium carbonate in water and dilute the solution to 1 litre.
Acetic acid reagent Dissolve 40 g of zinc sulphate ($ZnSO_4 \cdot 7H_2O$) and 70 g of KCl in 600 ml of water, slowly add 200 ml of glacial acetic acid and dilute the solution to 1 litre with water.
Buffer solution (*pH 4·6–4·8*) Dissolve 4·1 g of anhydrous sodium acetate in water, add 3 ml of glacial acetic acid and dilute the solution to 1 litre with water.
Sulphuric acid (3·7 N) Carefully add, with stirring, 10 ml of concentrated sulphuric acid to approximately 80 ml of water and dilute the solution with water to 100 ml.
Potassium iodide solution Add 50 g of KI to 100 ml of cold water and add one drop of 10 N NaOH. The solution must be freshly prepared.

Procedure

Introduce 5 g of flour and a teaspoonful of ignited sand into a

125-ml wide-mouthed bottle or flask and mix by rotation. Warm the mixture to 30 °C, add 46 ml of buffer solution (also at 30 °C) and rotate the container again until all the flour is in suspension. Place the container in a water-bath maintained at 30 °C for 1 h, shaking the mixture by rotation every 15 min. At the end of the 1 h period, add 2 ml of 3·7 N sulphuric acid, mix, add 2 ml of 12% sodium tungstate solution, mix again, allow the mixture to stand for 2 min, then filter it through a Whatman No. 4 filter-paper, discarding the first 10 drops. Immediately, pipette 5 ml of filtrate into a large test-tube, add by pipette exactly 10 ml of alkaline ferricyanide solution and immerse the test-tube in boiling water for 20 min with the surface of the liquid in the tube about 4 cm below that of the boiling water. Cool the tube under running water and wash the contents into a 100-ml conical flask with 25 ml of acetic acid reagent. Then add 1 ml of potassium iodide solution and 2 ml of starch solution, mix and back-titrate the mixture with 0·1 N sodium thiosulphate (x ml) until the blue colour has completely disappeared. (If the material in the tube is colourless instead of yellow after treatment in the boiling water bath and does not turn blue after adding KI, repeat the determination using less extract, e.g., 1 or 2·5 ml. In such cases, dilute the mixture to 5 ml with water and modify the foregoing calculation.)

$$y = \text{ml of } 0\cdot1 \text{ N ferricyanide reduced} = (10 - x) \text{ ml}$$

The maltose value, expressed as the percentage of maltose formed, can then be obtained by reference to *Table 8.1.* In interpreting values given in the literature, it must be borne in mind that in some countries, including the U.S.A., the maltose value is expressed as milligrams of maltose produced by 10 g of flour ($100 \times$ the U.K. value). To be suitable for bread, a flour should normally have a percentage maltose value of 2·0–3·5. Below this range, there is an inadequate supply of sugar produced diastatically in the dough and bread produced from it is likely to have a pale crust. The addition of malt flour causes a significant increase in the maltose value. A value above 3·5 suggests excessive α-amylase activity, as would be found in flour from sprouted wheat. Excessive α-amylase tends to give a sticky crumb in the bread obtained. Most bread flour has adequate β-amylase, which, however, only acts on damaged starch, the amount of which varies considerably[18]. In general, therefore, it is not absolutely certain that a high maltose value is synonymous with the presence of excessive α-amylase. A more direct measure of the α-amylase activity can be obtained from the Hagberg Falling Number test[19,20]

224

Table 8.1. MALTOSE VALUE OBTAINED USING BLISH AND SANDSTEDT'S METHOD[17]

$0·1$ N ferricyanide reduced, y ml	Maltose value, %	$0·1$ N ferricyanide reduced, y ml	Maltose value, %	$0·1$ N ferricyanide reduced, y ml	Maltose value, %
0·1	0·05	3·4	1·71	6·7	3·79
0·2	0·10	3·5	1·76	6·8	3·85
0·3	0·15	3·6	1·82	6·9	3·92
0·4	0·20	3·7	1·88	7·0	3·98
0·5	0·25	3·8	1·95	7·1	4·06
0·6	0·31	3·9	2·01	7·2	4·12
0·7	0·36	4·0	2·07	7·3	4·18
0·8	0·41	4·1	2·13	7·4	4·25
0·9	0·46	4·2	2·18	7·5	4·31
1·0	0·51	4·3	2·25	7·6	4·38
1·1	0·56	4·4	2·31	7·7	4·45
1·2	0·60	4·5	2·37	7·8	4·51
1·3	0·65	4·6	2·44	7·9	4·58
1·4	0·71	4·7	2·51	8·0	4·65
1·5	0·76	4·8	2·57	8·1	4·72
1·6	0·80	4·9	2·64	8·2	4·78
1·7	0·85	5·0	2·70	8·3	4·85
1·8	0·90	5·1	2·76	8·4	4·92
1·9	0·96	5·2	2·82	8·5	4·99
2·0	1·01	5·3	2·88	8·6	5·05
2·1	1·06	5·4	2·95	8·7	5·12
2·2	1·11	5·5	3·02	8·8	5·19
2·3	1·16	5·6	3·08	8·9	5·27
2·4	1·21	5·7	3·15	9·0	5·34
2·5	1·26	5·8	3·22	9·1	5·42
2·6	1·30	5·9	3·28	9·2	5·50
2·7	1·35	6·0	3·34	9·3	5·58
2·8	1·40	6·1	3·41	9·4	5·68
2·9	1·45	6·2	3·47	9·5	5·78
3·0	1·51	6·3	3·53	9·6	5·88
3·1	1·56	6·4	3·60	9·7	5·98
3·2	1·61	6·5	3·67	9·8	6·08
3·3	1·66	6·6	3·73	9·9	6·18

CHALK

The Bread and Flour Regulations 1963[1] require all flours other than wholemeal, wheat malt flour and self-raising flour that contain not less than 0·2 % of Ca (due in practice to the presence of acid calcium phosphate as the acid ingredient) to contain 235–390 mg of chalk per 100 g (ref. 2). The amount present is usually estimated by ashing 25 g of flour, precipitating the calcium as oxalate and titrating with permanganate (*Method 4.3*).

1 ml of 0·05 N potassium permanganate

$\equiv$ 1 mg of Ca $\equiv$ 2·50 mg of $CaCO_3$

$\equiv$ 2·58 mg of chalk of minimum BP quality (min. 97 % $CaCO_3$)

An allowance for the natural calcium present in the flour must be made by subtracting an amount varying from 15 mg (white flour) to 30 mg (wholemeal) of Ca per 100 g. Other methods for estimating chalk have been described based on back titration[21, 22], weighing the CO_2 absorbed in a gravimetric train[21], gasometrically in a modified Chittick apparatus[23] and absorptiometrically using chloranilic acid[24]. The application of the Chittick method to self-raising flours, which also contain bicarbonate, is given on p. 235.

IRON

The Bread and Flour Regulations 1963[1] require all flour to contain at least 1·65 mg of Fe per 100 g[2]. If fortification is necessary, the addition must be made in the form of reduced iron or ferric ammonium citrate. Iron is determined colorimetrically on a solution of the ash using either 2,2'-dipyridyl (see *Method 4.5*, p. 105) or *o*-phenanthroline[25].

THIAMINE

The Bread and Flour Regulations 1963[1] require all flour to contain at least 0·24 mg of vitamin B_1 (thiamine or aneurine) per 100 g[2]. Most of the methods for determining thiamine in flour are based on the thiochrome reaction of Jansen.

Principle of method

The thiamine is extracted with dilute HCl and, after oxidation using alkaline ferricyanide, the blue fluorescence is compared fluorimetrically[26, 27] or visually with that produced by a standard thiamine solution similarly treated. The method described below can be performed without recourse to special equipment other than an ultraviolet lamp[28].

226

METHOD 8.8. ESTIMATION OF THIAMINE IN FLOUR

Reagents

Standard aneurine (*stock solution*) 20 μg of aneurine hydrochloride per millilitre of 0·01 N HCl containing 25% alcohol. Store in a refrigerator.

Dilute standard aneurine solution Dilute the stock solution on the day of the test with 0·01 N HCl so that 1·5 ml contains 4 μg of aneurine hydrochloride.

1 International Unit of vitamin B_1 $\equiv$

$$3 \text{ μg of aneurine hydrochloride}$$

Procedure

Cream 2 g of the flour with 25 ml of 2% HCl, allow the mixture to stand overnight, then shake and centrifuge it. Transfer 5 ml of the clear liquid to a 15-ml graduated centrifuge tube, add one drop of 10% HCl and 5 ml of isobutanol. Shake the tube for at least 1 min, centrifuge and note the volume of the aqueous layer (containing the vitamin).

Into a 25-ml stoppered bottle, A, place 1·5 ml of the aqueous layer, add 2 ml of methanol, 1 ml of 20% NaOH, and then, within 30 s, 0·15 ml of 5% potassium ferricyanide solution, shaking the mixture thoroughly after each addition. Add 0·25 ml of hydrogen peroxide (20 volume) and allow the solution to stand for 1 min. Add 10 ml of isobutanol, shake the mixture vigorously for 2 min and allow it to separate. In a second bottle, B, place 1·5 ml of the aqueous layer from the centrifuge tube and follow the same procedure but omit the ferricyanide. Into a third bottle, C, place 1·5 ml of dilute standard aneurine solution and follow the same procedure as with bottle A. To compare the fluorescence of the thiochrome produced, draw off the aqueous layer from each bottle with a pipette, wash the isobutanol with 3 ml of water, remove the water, and add 2 ml of ethanol to each. Measure 10 ml of the clear extract from bottles A and B into separate tubes. Compare the fluorescence in a darkened room close to an ultraviolet lamp. Using a semi-micro burette, add volumes of the standard in C to the blank in B until the fluorescence observed matches that of the solution in A. At the last matching, add a volume of isobutanol to the sample in A equal to that of the standard added to the blank in B.

Calculation

If x ml is the volume of standard isobutanol extract in C required and r_1 and r_2 ml are the volumes of the extracts before and after the preliminary washing with isobutanol, then the aneurine content of the sample is:

$$\frac{4x}{10} \times \frac{25}{1 \cdot 5 \times 2} \times \frac{r_2}{r_1} = \frac{10 r_2 x}{3 r_1} \mu g/g$$

(minimum 1·60 mg/100 g)

The *nicotinic acid* in flour can be estimated by microbiological assay[29] or chemically[30].

BAKING POWDER

Baking powder is used for aeration in baking and contains an acidic ingredient, sodium bicarbonate, and a cereal or starch filler. Although cream of tartar (potassium hydrogen tartrate) and tartaric acid have been used, acid calcium phosphate (ACP) and acid sodium pyrophosphate (ASP) are now by far the most commonly used commercially as acid ingredients. In general, ACP reacts more rapidly with the bicarbonate than ASP, during storage, product preparation and cooking. As the individual ingredients store well, however, it is often preferable in the factory to weigh them out separately for the process. For convenience in weighing out, the acid ingredients are obtainable as *cream powders*, i.e. powders in which the acid is diluted with starch. The dilution is such that the acid phosphate in 2 parts of cream powder is equivalent to 1 part of $NaHCO_3$. The available and residual carbon dioxide in baking powder is controlled by law[31] (see p. 230).

RAW MATERIALS

In general, attention must be paid to suitable granularity and low moisture content so as to reduce caking and the rate of loss of carbon dioxide in the product.

FILLER

Examine the filler microscopically for the type of starch by mounting it in water and comparing it against type specimens. Also, estimate the moisture content by drying a sample at 155 °C (Carter–Simon method, *Method 2.2d*). The moisture content should not exceed about 9%.

SODIUM BICARBONATE

Check the alkalinity of a sample by dissolving about 1 g, accurately weighed, in 20 ml of cold water and titrating the solution with 0·5 N HCl (V ml) using methyl orange as indicator.

$$\text{NaHCO}_3 \ (\%) = \frac{V \times 0{\cdot}0420 \times 100}{\text{Weight taken}}$$

The NaHCO_3 content determined should preferably fall within the range 99·0–101·0%. Limit tests for impurities such as arsenic and lead are described in the BP monograph for sodium bicarbonate.

ACID PHOSPHATES AND CREAM POWDERS

METHOD 8.9. DETERMINATION OF THE NEUTRALISING VALUE OF ACID PHOSPHATES

ASP Weigh exactly 0·84 g of sample into a 400-ml beaker flask and add 20 g of NaCl and 25 ml of water. Stir the mixture with a flat-ended rod for 4 min and add from a burette 90·0 ml of 0·1 N NaOH. Stir the solution, add phenolphthalein and titrate with 0·2 N HCl (A ml).

Neutralising value = parts by weight of pure NaHCO_3
                           equivalent to 100 parts acid phosphate
                           sample
           = $90 - 2A$

ACP Weigh exactly 0·84 g of sample into a 400-ml beaker flask, add 25 ml of cold water, stir the mixture and then add from a burette 90·0 ml of 0·1 N NaOH. Boil the mixture for 1 min, add phenolphthalein and titrate while hot with 0·2 N HCl. Calculate the neutralising value as for ASP.

In devising recipes, the following rules should be borne in mind:

(*a*) The bicarbonate should be in *slight* excess of the acid ingredient. For calculations typical equivalents are:

229

1 part of sodium bicarbonate $\equiv$
$$1{\cdot}36 \text{ parts of commercial ASP}$$
1 part of sodium bicarbonate $\equiv$
$$1{\cdot}25 \text{ parts of commercial ACP}$$

(b) The total CO_2 in a baking powder intended for retail sale should not be less than about 12%.

(c) Statutory requirements (see Section on 'Product' below).

Acid phosphates may also require examination for impurities such as fluorine[32, 33], arsenic and lead (see Chapter 4) and sulphate.

PRODUCT

At the time of writing, baking powder is required to contain at least 8% of available CO_2 (or 6% for golden raising powder, which contains a yellow dye) and not more than $1{\cdot}5\%$ of residual CO_2. The first part of the estimation of each of these components is prescribed in the regulations[31]. Excessive residual CO_2 in a baking powder gives rise to a soda-taste in cakes made from it.

$$\text{Total sodium bicarbonate} \xrightarrow{+H_2SO_4} \text{Total CO}_2 = T$$

The available CO_2 is that produced when water is added to the sample and heated to boiling:

$$\text{Total acid ingredient} + \text{Equivalent bicarbonate} \xrightarrow{H_2O}$$
$$\text{Available CO}_2 = A$$

In the analysis, the sample is evaporated with water, leaving any unreacted or residual bicarbonate (in excess of the equivalent of acid present). Then, on adding mineral acid to the residue:

$$\begin{array}{c}\text{Residue containing unreacted} \\ \text{bicarbonate in excess of the} \\ \text{equivalent acid in the sample}\end{array} \xrightarrow{+H_2SO_4} \text{Residual CO}_2 = R$$

For the purposes of checking samples for compliance with regulations, T and R are estimated by the published procedures, and:

$$A = T - R$$

METHOD 8.10. ESTIMATION OF THE TOTAL RESIDUAL AND AVAILABLE CARBON DIOXIDE IN BAKING POWDER AND GOLDEN RAISING POWDER USING THE CHITTICK APPARATUS (CF., REGULATIONS[31])

Chittick apparatus (Figure 8.3)

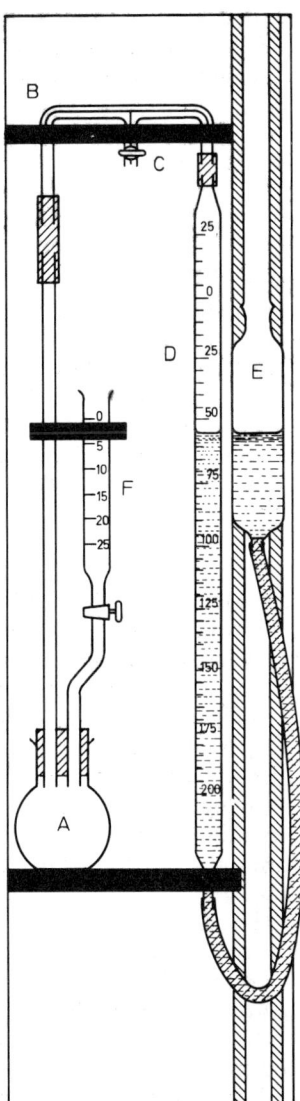

Figure 8.3. Chittick apparatus for determining carbon dioxide

The wide-mouthed 250-ml flask, A, connected to the T-tube, B, and fitted with a stopcock, C, is connected to a gas-measuring burette, D, with marks at -25 ml ($=$ volume of sulphuric acid used), 0 ml and $+200$ ml. The burette is connected by means of rubber tubing to the 300-ml levelling bulb, E. Flask A is fitted with a two-hole rubber stopper, through one hole of which passes a 25-ml burette, F, and through the other a glass tube of the same diameter as T-tube B.

N.B. Secure all rubber tubing joints with strong copper wire. Grease the glass taps.

Reagents

Sulphuric acid solution, G To 1 litre of water carefully add 200 ml of concentrated sulphuric acid, stirring continuously. Cool the solution.

Displacement solution, H Dissolve 100 g of NaCl in 350 ml of water and add 1 g of sodium bicarbonate and 2 ml of 0·5% aqueous methyl orange solution. Then add, with stirring, just enough of sulphuric acid solution G to make the solution acidic (definite pink colour).

Procedure for total CO_2

Check all joints of the Chittick apparatus. By means of levelling bulb E, introduce a suitable volume of displacement solution H into burette D. Close the tap on burette A. Weigh out 0·5–1·0 g of sample into flask A, add several large glass balls and connect A to the apparatus. Open tap C and move E so as to bring the liquid in burette D close to the upper 25-ml mark. Allow 5 min to elapse to enable the apparatus to equilibrate with the room temperature before bringing the level exactly on to the upper 25-ml mark. Then close C and lower E to reduce the pressure in the system (and at all times during the subsequent reaction keep the level in E lower than that in D). Fill burette F to the zero mark with sulphuric acid solution G and add exactly 25 ml to flask A. Shake the flask well, holding it by the rim to avoid heating. Allow flask A to stand for 2 min, then re-shake it. Then allow it to stand for about 10 min to secure temperature equilibrium and move E so that the level of liquid corresponds exactly with that in D. Read off the volume of CO_2 produced (the subtraction for the 25 ml of solution G added is allowed for if the apparatus shown in *Figure 8.3* is used). Note the atmospheric temperature and pressure, correct the CO_2 volume to NTP, calculate the weight of gas produced and express it as a percentage of the sample.

Density of CO_2 = 1·976 g/l at 0 °C and 760 mm

N.B. The apparatus should be standardised from time to time with A.R. sodium bicarbonate.

Calculation

Room temperature $= 21\,^{\circ}C = 294\,^{\circ}K$

Atmospheric pressure $= 765$ mm

Weight of baking powder taken $= 0.550$ g

Volume of CO_2 produced $= 45.0$ ml

Equivalent volume of CO_2 at NTP $= 45.0 \times \dfrac{765}{760} \times \dfrac{273}{294}$

$\qquad\qquad\qquad\qquad = 42.1$ ml

Weight of CO_2 produced $= \dfrac{42.1 \times 1.976}{1000}$

$\qquad\qquad\qquad\qquad = 0.0831$ g

Total $CO_2 = \dfrac{0.0831}{0.550} \times 100$

$\qquad\qquad = 15.1\%$

Procedure for residual CO_2

Weigh out 2 g of sample into the wide-mouthed 250-ml flask, A, add 25 ml of water and evaporate the solution to dryness on a boiling water bath. To the residue add a further 25 ml of water, break up the solid with a glass rod and evaporate the mixture to dryness again. Continue as described for the estimation of the total CO_2 (including the calculation) commencing at 'add several large glass balls. . . .'.

Available CO_2 (%) $= \%$ Total $CO_2 - \%$ Residual CO_2

Miscellaneous methods

A quick check in production control, which confirms that both the acid and bicarbonate are present, involves spreading out some of the sample and pressing a filter-paper soaked in universal indicator down on to it. The acid and alkaline ingredients give differently coloured spots on the paper.

The total and available carbon dioxide levels in baking powder diminish during storage. Methods for estimating the original composition of deteriorated samples of baking powder have been discussed by the author[34]. Although in industrial control the fluorine content (maximum 15 ppm) is best carried out on the acid phosphate (maximum 30 ppm calculated on the acid phosphate) when received as a raw material, a check on the product may also be necessary[32, 33].

SELF-RAISING FLOUR

Self-raising flour is essentially flour that contains a small amount of baking powder. The flour should be a weak flour with a relatively low maltose value and of suitable granularity and moisture content[35]. An excessive moisture content encourages loss of carbon dioxide during storage. The flour is mixed with about 1.15% of sodium bicarbonate ($\equiv 0.6\%$ of CO_2) and its equivalent of ACP (1.44%) or ASP (1.56%). The examination of each of these raising materials is discussed earlier in this Chapter.

Self-raising flour must comply with the statutory requirements for flour [given in previous Sections in this Chapter except that no chalk addition is necessary if the ACP present supplies at least 0.2% of Ca (1.44% of commercial ACP contains 0.26% of Ca)]. The Food Standards (Self-Raising Flour) Order 1946 prescribes a minimum of 0.40% of available CO_2. Also, the first part of the estimation of the total and residual CO_2 is prescribed in this Order. To keep losses of aerating power and product weight to a minimum, self-raising flour should be stirred at an R.H. of 55–75% and below 65°F.

In routine industrial control, large numbers of samples can be checked by estimating the total CO_2 and confirming that both acid and alkaline ingredients are present by applying the test using universal indicator (see p. 233). The method for estimating the carbon dioxide due to bicarbonate requires some modification if added chalk is present. Legally, self-raising flour must not contain more than 3 ppm of fluorine[32, 33]. The modified preparation of the paste for estimating the Kent-Jones and Martin Flour Grade Figure in self-raising flour is included in *Method 8.5*.

METHOD 8.11. ESTIMATION OF THE TOTAL, RESIDUAL AND AVAILABLE CARBON DIOXIDE IN SELF-RAISING FLOUR USING THE CHITTICK APPARATUS IN THE ABSENCE OF CHALK

Apparatus and Reagents—See *Method 8.10*.

Procedure for total CO_2

Proceed as described for baking powder (*Method 8.10*) using a 10-g sample.

Procedure for residual CO_2

Weigh out 5 g of sample into flask A and mix it into a smooth paste with water. Then mix the paste with about 100 ml of water and

234

place the flask in boiling water for 30 min, stirring very frequently. Then boil the liquid for 3 min, stirring vigorously throughout. Cool the mixture thoroughly and continue as described in *Method 8.10* commencing at 'add several large glass balls. . . .'

$$\text{Available } CO_2 (\%) = \% \text{ Total } CO_2 - \text{Residual } CO_2$$

METHOD 8.12. ESTIMATION OF THE TOTAL CARBON DIOXIDE IN SELF-RAISING FLOUR IN THE PRESENCE OF CRETA

Principle of method[36]

The addition of a 2·5% solution of acid sodium pyrophosphate to self-raising flour containing chalk causes the CO_2 to be released from the bicarbonate, but not from the chalk. Dilute sulphuric acid reacts with both.

Apparatus—See *Method 8.10* and *Figure 8.3*.

Reagents

Sulphuric acid solution, G See *Method 8.10*.
Displacement solution, H See *Method 8.10*.
ASP solution Dissolve 25 g of acid sodium pyrophosphate in cold water and dilute to 1 litre.

Procedure for CO_2 derived from sodium bicarbonate

Proceed as for the estimation of the total CO_2 in baking powder (*Method 8.10*) with the following modifications:

(*a*) Weigh out 17 g of self-raising flour.

(*b*) Instead of adding 25 ml of sulphuric acid solution G from burette F, add 45 ml of the ASP solution in successive volumes of 25 and 20 ml.

(*c*) In reading off the volume of CO_2 produced from the sample, allow for the 45 (25 + 20) ml of ASP solution added.

(*d*) Repeat the procedure without the sample to give the reagent blank. Subtract this volume from that obtained with the sample to give that due to the bicarbonate present.

(*e*) Tests should also be made on known mixtures and an appropriate factor applied.

Procedure for CO_2 derived from sodium bicarbonate and chalk

Proceed as in *Method 8.11* for total CO_2 using dilute sulphuric acid. The difference between the two results gives the carbon dioxide due to chalk ($CaCO_3/CO_2 = 2\cdot27$).

REFERENCES

1. *The Bread and Flour Regulations 1963 (SI 1963 No. 1435)*, H.M.S.O., London
2. *Food Standards Committee Report on Bread and Flour*, H.M.S.O., London (1960)
3. KENT-JONES, D. W. and AMOS, A. J., *Modern Cereal Chemistry*, 6th edn, Food Trade Press, London (1967)
4. *The Approved Methods of the AACC* (American Association of Cereal Chemists, Minnesota)
5. VAN SOEST, P. J., *J. Ass. off. agric. Chem.*, **46**, 825 (1963)
6. TKACHUK, R., *Cereal Chem.*, **43**, 207 (1966)
7. TKACHUK, R., *Cereal Chem.*, **43**, 223 (1966)
8. UDY, D. C., *Cereal Chem.*, **33**, 190 (1956)
9. UDY, D. C., *Cereal Chem.*, **34**, 389 (1957)
10. FEINSTEIN, L. and HART, J. R., *Cereal Chem.*, **36**, 191 (1959)
11. MORRISON, W. R., *J. Sci. Fd Agric.*, **14**, 245 (1963)
12. MORRISON, W. R., *J. Sci. Fd Agric.*, **14**, 870 (1963)
13. KENT-JONES, D. W., AMOS, A. J., MARTIN, W., SCOTT, R. A. and ELIAS, D. G., *Chemy Ind.*, 1490 (1956)
14. GILLIS, J. A., *Cereal Sci. Today*, **8**, No. 2, 40 (1963)
15. KENT-JONES, D. W., RICHARDSON, E. G. and SPALDING, R. C., *J. Soc. chem. Ind., Lond.*, **58**, 261 (1939)
16. STEVENS, D. J., *J. Sci. Fd Agric.*, **14**, 405 (1963)
17. BLISH, M. J. and SANDSTEDT, R. M., *Cereal Chem.*, **10**, 189 (1933)
18. GREER, E. N. and STEWART, B. A., *J. Sci. Fd Agric.*, **40**, 248 (1959)
19. HAGBERG, S., *Cereal Chem.*, **37**, 218 (1960)
20. HAGBERG, S., *Cereal Chem.*, **38**, 202 (1961)
21. GREER, E. N., MOUNFIELD, J. D. and PRINGLE, W. J. S., *Analyst, Lond.*, **67**, 352 (1942)
22. HARTLEY, A. W. and GREEN, A., *Analyst, Lond.*, **68**, 142 (1943)
23. FRASER, J. R. and WESTON, R. E., *Analyst, Lond.*, **75**, 402 (1950)
24. SAWYER, R., TYLER, J. F. C. and WESTON, R. E., *Analyst, Lond.*, **81**, 362 (1956)
25. PRINGLE, W. J. S., *Analyst, Lond.*, **71**, 490 (1946)
26. RIDYARD, H. N., *Analyst, Lond.*, **74**, 18 (1949)
27. SOCIETY OF PUBLIC ANALYSTS, *Analyst, Lond.*, **76**, 127 (1951)
28. MEDICAL RESEARCH COUNCIL, *Biochem. J.*, **37**, 433 (1943)
29. SOCIETY OF PUBLIC ANALYSTS, *Analyst, Lond.*, **71**, 397 (1946)
30. DENNIS, P. O. and REES, H. G., *Analyst, Lond.*, **74**, 481 (1949)
31. *The Food Standards (Baking Powder and Golden Raising Powder) Order 1944 (SR & O, 1944 No. 46; 1946 No. 157)*, H.M.S.O., London
32. SOCIETY OF PUBLIC ANALYSTS, *Analyst, Lond.*, **69**, 243 (1944)
33. SOCIETY OF PUBLIC ANALYSTS, *Analyst, Lond.*, **70**, 442 (1945)
34. PEARSON, D., *The Chemical Analysis of Foods*, 6th edn, Churchill, London (1970)
35. HARTLEY, A. W. and GREEN, A., *Analyst, Lond.*, **70**, 211 (1945)
36. STEPHENSON, W. H. and HARTLEY, A. W., *Analyst, Lond.*, **80**, 461 (1955)

9

SUGAR AND FRUIT PRODUCTS

Sugar—Jam—Canned Fruits—Pickles and Sauces—
Vinegar—Tomato Purée—Fruit Juices

SUGAR

Refined sucrose sugar is an important raw material in factories making preserves, confectionery, fruit products, etc. The main analytical control methods used for assessing its quality are as follows.

MOISTURE

Dry at 105 °C, or preferably in a vacuum oven at 60–70 °C (p. 31). The moisture content of white sugar should be less than 0.1%.

ASH

Determine the sulphated ash (*Method 2.7e*) or by means of the electrical conductivity[1]. The sulphated ash of white sugar is less than 0.02%.

REDUCING SUGARS

Determine by Lane and Eynon's method (*Method 2.8a*) or the EDTA back-titration method below.

METHOD 9.1. DETERMINATION OF INVERT SUGAR IN REFINED SUGAR USING EDTA[2]

Principle of method

The sugar is heated with an alkaline copper solution and the degree

237

of reduction is assessed by titrating the excess of unreacted copper with EDTA.

Reagents

EDTA solution (0·005 N) Dissolve 0·930 g of the disodium salt of ethylenediaminetetra-acetic acid in water and dilute the solution to 1 litre with water.

Alkaline copper solution Mix 40 ml of 1 N NaOH with 600 ml of water, add 25 g of sodium carbonate and 25 g of potassium sodium tartrate and shake the mixture to dissolve the solids. Then add a copper sulphate solution containing exactly 6·000 g of $CuSO_4·5H_2O$ in about 100 ml of water and dilute to exactly 1 litre with water.

Murexide indicator Grind together 0·5 g of murexide, 0·15 g of methylene blue and 40 g of NaCl.

Procedure

Weigh out $5 \pm 0·005$ g of the sugar sample into a boiling-tube and dissolve it in 5 ml of water. Add exactly 2 ml of alkaline copper solution, mix and immerse the tube in boiling water for 5 min. Cool the tube immediately in cold water. Wash the solution into a white porcelain dish with water, add 0·1 g of murexide indicator and titrate with 0·005 N EDTA. The colour change is first from green to grey and then to purple at the end-point. The titration at the end-point should not be too prolonged. The amount of invert sugar can be obtained by reference to a graph relating the percentage directly to the EDTA titrations[2]. The graph is linear up to about 0·02% of invert sugar and an equation such as the following can be used for most samples of white sugar:

$$\text{Invert sugar } (\%) = 0·0199 - [0·0015 \times \text{Titration } (0·005 \text{ N EDTA})]$$

For good keeping properties, the invert sugar content of white sugar should be less than 0·1%.

POLARISATION

Determine the rotation of a 10% or 26% solution in a tube 2 dm in length and calculate as sucrose (p. 68).

INSOLUBLE MATTER

Determine by the method of Hibbert and Phillipson[3].

pH VALUE

Determine on a solution of 60° Brix (60% sugar by weight). The pH should preferably be just on the acidic side of 7.

SULPHUR DIOXIDE

Sulphur dioxide can be determined by one of the distillation methods described in Chapter 3. Trace amounts can be determined colorimetrically by the following method using bleached rosaniline and formaldehyde.

METHOD 9.2. COLORIMETRIC ESTIMATION OF SULPHUR DIOXIDE IN WHITE SUGAR USING ROSANILINE[4]

The following method has been developed from the bleached rosaniline colorimetric procedures of Steigmann[5] and West and Gaeke[6].

Reagents

(A) *Saturated aqueous rosaniline hydrochloride*—Suspend 1 g of rosaniline hydrochloride in 100 ml of water and heat the mixture to about 50 °C. Allow it to cool with shaking, allow it to stand for 48 h with occasional shaking, then filter it.

(B) *Bleached rosaniline*—To 4 ml of A add 6 ml of concentrated HCl, mix and dilute the solution to 100 ml with water. Allow the solution to stand for at least 1 h before using it.

(C) *0·2% Formaldehyde solution*—Dilute 5 ml of formalin (40% formaldehyde) to 1 litre with water.

(D) *Sodium sulphite standard solution*—Dissolve 100 g of sulphite-free sucrose in water and transfer to a 1-litre volumetric flask. Then add 0·5 g of sodium sulphite ($Na_2SO_3 \cdot 7H_2O$), also dissolved in water, mix and dilute the solution to the mark with water. To standardise the solution, add 50 ml of 0·1 N iodine solution to 100 ml of the sulphite–sucrose solution, allow the mixture to stand for 5 min, then add 1 ml of concentrated HCl and titrate the excess iodine with 0·1 N sodium thiosulphate using starch as indicator.

$$1 \text{ ml of } 0\cdot1 \text{ N thiosulphate solution} \equiv 0\cdot01261 \text{ g of } Na_2SO_3\cdot7H_2O$$
$$\equiv 0\cdot003203 \text{ g of } SO_2).$$

Procedure

Prepare an aqueous solution of the sugar sample so that it contains 0·5–30 μg of SO_2 per 10 ml and is approximately 0·004 N with respect to sodium hydroxide. This can usually be achieved by using 10–20 g of sucrose and 4 ml of 0·1 N NaOH per 100 ml. To a clean test-tube add by pipette 10 ml of sugar solution, 2 ml of bleached rosaniline (*B*) and then 2 ml of 0·2% formaldehyde (*C*). Mix and, after allowing the mixture to stand for 25–35 min, measure the optical density at 560 nm and compare the result against the calibration curve. Express the amount of sulphur dioxide present as ppm of SO_2 in the sample.

Preparation of calibration curve

Dilute the sodium sulphite standard solution (*D*) with a solution containing 10% of sulphite-free sucrose so that 1 ml contains 50 μg of SO_2 (solution *E*). Then to a series of 100-ml volumetric flasks add by pipette 0·0, 1·0, 2·0, 3·0, 4·0, 5·0 and 6·0 ml of *E* and 4 ml of 0·1 N NaOH. Dilute each to 100 ml with 10% sucrose (sulphite-free) solution. Take 10 ml of each solution in a test-tube and continue as for the sample by adding 2 ml of bleached rosaniline (*B*), etc. Prepare the calibration curve by plotting the optical density at 560 nm against the concentration as μg per 10 ml of solution taken.

The maximum concentration of sulphur dioxide allowed in sugar is 70 ppm (as SO_2).

BULK DENSITY

The values obtained for bulk density vary according to the technique and apparatus used. The estimation of bulk density is particularly important with refined granulated sugar for which the method described below is more specifically designed.

METHOD 9.3. ESTIMATION OF BULK DENSITY OF SUGAR BY MEADE'S METHOD

Principle of method

Using a standardised technique, the sugar sample is weighed after falling from a funnel into a cylinder of known volume. The first procedure described measures the 'as poured' or 'unsettled' bulk density.

Apparatus

A stemless metal funnel (capacity approximately 1 litre) is used with an apex angle of the cone of 60° and a circular opening at the bottom of 1 cm diameter. The funnel is closed with a sliding gate and supported on legs so the bottom exit is 3 cm above the top of a metal cylinder (4–5 cm diameter, 625 ml capacity).

Procedure

Check the volume of the cylinder by filling it with water and weighing it. Close the gate on the funnel and support it over the dried cylinder. Place in the funnel sufficient sugar (650–750 ml) to more than fill the cylinder. Open the gate, allow the cylinder to fill to overflowing and level off the sugar with a spatula. Avoid tapping or knocking the cylinder at this stage. Weigh the cylinder when full with sugar. Return the sugar to the funnel, repeat the operation and weigh the cylinder to the nearest gram. By subtraction, obtain the weight of sugar (M g) of volume V ml.

$$\text{Bulk density (g/ml)} = \frac{M}{V}$$

$$\text{Bulk density (lb/ft}^3) = \frac{62 \cdot 5 \, M}{V}$$

To obtain the bulk density after compacting, tap the cylinder by gently dropping it on to a wooden surface from a height of 1·2 cm once per second, adding more sugar from the funnel until the cylinder remains full after tapping. Reweigh the sugar and calculate the 'settled' bulk density from the above equations.

COLOUR

The colour of sugar is normally assessed from the optical densities of a concentrated solution (ICUMSA method).

METHOD 9.4. ASSESSMENT OF THE COLOUR OF SUGAR BY MEANS OF A SPECTROPHOTOMETER

Prepare a 50% solution of sugar by adding 100 ml of water to 100 g of sugar and stir to dissolve. (With liquid sugars, dilute with water to 50% solids as determined by a refractomer.) Filter the mixture and measure the optical density of the solution at 420 nm (D_{420}) and 720 nm (D_{720}), preferably in a 10-cm cell against water as reference standard (cf., ICUMSA).

$$\text{Colour Index} = \frac{1000(D_{420} - D_{720})}{lc}$$

where l = length of cell (cm) = 10 cm in the above method, and c = concentration of sugar (g/ml).

The Colour Index of refined white sugar should be less than 10.

JAM

Jam is prepared by boiling together prepared fruit (usually sulphited pulp), sugar syrup (approximately 66% soluble solids), glucose syrup (82%, 43° Baumé) and water[7]. The proportion of glucose sugars used should not exceed 15% of the total sugars (both calculated as dry sugars in the product). When the fruit is naturally deficient in pectin strength and/or acid, the deficiency is made up by adding solutions of pectin or acid (citric, tartaric or malic) near the end of the boiling. Solutions of colouring agents are also often added, particularly if the fruit source is sulphited pulp. After partially cooling (partly to 'stop' further inversion of sucrose) and stirring (to keep the mixture homogeneous), the preserve is filled into washed jars, which are hermetically sealed. Each batch is checked refractometrically for soluble solids, the level of which should be at least 66–5% to ensure that the legal minimum of 65% is complied with (or 70% as measured to ensure a level of 68·5% if the jars are not hermetically sealed). Boiling in a vacuum pan reduces caramelisation, but owing to the lower degree of inversion at the lower temperature it is often necessary to replace some of the sucrose syrup with invert syrup to ensure that the final product contains over, say, 25% of invert sugar.

The production of jelly jam involves boiling the fruit with water and boiling the filtered extract with the sugar. With marmalade, the centres ('dummies' or 'hearts') and the cut peel are first boiled separately in water. The peel is thus softened so that it is able to absorb sugar more easily in the main boil.

The overall aim in jam manufacture is to obtain a product with good appearance and taste, which will withstand handling and transport, as a result of the formation of a strong gel owing to the presence of pectin, sugar and acid. There should be present sufficient sugar to prevent the growth of micro-organisms and the correct proportion of reducing sugars to prevent crystal formation.

INGREDIENTS

SUGAR

For jam, the solid sucrose used (as such or in syrup) should be white sugar giving a polarisation value of at least 99·8% (Section 2.8) and the sulphated ash (*Method 2.7e*) should not exceed 0·025%. The pH (Section 2.10) should be between 6·4 and 7·2.

Invert syrup is prepared by heating sucrose solution with HCl to 200 °F for 15–20 min, after which the acidity is neutralised with bicarbonate. The soluble solids content is checked with a refractometer (*Tables 2.4*, p. 59, and *2.4A*, p. 60; cf., *Table 9.1*, p. 248).

FRUIT

In view of checks to be made on the product, it is useful to estimate from time to time the potassium and insoluble solids on the fresh fruit arriving at the factory. Apart from the fresh material, the fruit used is preserved by freezing, heat sterilisation or the addition of a chemical preservative. Many fruits are heated with water to facilitate the release of pectin before covering them with sulphurous acid, but to retain the firmness in delicate fruits such as strawberries a cold process is used. The sulphurous acid (containing 6% of SO_2), the strength of which is checked from its density, is added to the water–fruit mixture to produce a final concentration of 1500–3000 ppm of SO_2, depending on the particular fruit. The containers should be marked with all relevant information, including dates of inspection. The soluble solids content is checked with a refractometer and the SO_2 content by the following direct titration method.

METHOD 9.5. DETERMINATION OF SULPHUR DIOXIDE IN FRUIT PULPS

Remove some of the liquid from the container with a large pipette, filter it if necessary and transfer it to a burette. To a conical flask add about 100 ml of water, 10 ml of 0·05 N iodine and starch solution and titrate with the fruit pulp liquor until the blue colour is completely discharged.

$$SO_2 \text{ (ppm)} = \frac{10 \times 0.05 \times 3.2 \times 10^4}{\text{Titre (ml of fruit pulp liquid)}}$$

$$= \frac{1.6 \times 10^4}{\text{Titre}}$$

A field test using standard tablets has also been devised.[8]

PECTIN

Pectose, from which pectin is derived, is abundant in citrus fruits, apples, pears, gooseberries and plums. The chief source for the manufacture of commercial pectin is apple pomace, i.e., the material remaining on the cider presses after the juice has been pressed out. The pomace is treated with hot dilute acid and, after clarifying, the filtered extract is concentrated to 3–4% pectin in a vacuum evaporator. The concentrate is heat-sterilised and sulphur dioxide (maximum 250 ppm) added. Powdered pectin, which is also produced, has to be mixed with concentrated sucrose syrup and stirred with boiling water before use.

The grade of a pectin can be defined as the ratio of total soluble solids to pectin in a jelly of standard strength prepared in a standard manner with a total soluble solids content between 70 and 71%[9]. In practice, it may be considered as the weight of sugar in pounds such that 1 lb of pectin will set to a jelly of standard strength under standard conditions, i.e., 68·5% soluble solids and pH 3·3 (cf., Olliver, Wade and Dent[10]).

The jams that require additional pectin for a good set are those which contain fruits that are either low in pectin or (often more important) contain pectin of low quality, e.g., cherry, fig, peach, pear, pineapple, raspberry, rhubarb and strawberry (see also Molyneux[11]).

Although the pectin content can be estimated as calcium pectate, the commercial quality can more conveniently be assessed by mixing 45 ml of pectin extract with 10 ml of alcohol or acetone in a beaker. If placed in ice, a good quality product gives a well set jelly within about 1 h. Manufacturers issue instructions as to the method of using each grade of pectin they supply.

Pectin concentrates may need checking for trace elements (Chapter 4), for which the following maxima apply (ppm):

	Statutory limits		FSC recommended limits
	Lead	Arsenic	Copper
Liquid pectin	10	2	30
Solid pectin	50	5	300

ACIDS AND BUFFERS

Depending on the particular fruit, it may be necessary to control the pH of the product and the degree of inversion of sucrose by the addition of acid or alkali ('buffer salts'). The citric acid, tartaric acid or sodium citrate or bicarbonate to be used should be titrated for quality. The corresponding BP monographs are useful guides to the types of test necessary. The only acids that can legally be added to jam in the U.K. are citric, tartaric and malic acids. Formerly, lactic acid was preferred by many jam manufacturers.

COLOURINGS

Artificial colouring is normally used in jams made from sulphited pulp. In the preparation of pulps, the colour of the fruit tends to be bleached by the sulphur dioxide. With red-coloured fruits such as red currants, loganberries, red plums and raspberries, the boiling of the jam causes a restoration of most of the colour. With strawberries, however, the colour restoration is less apparent and artificial colouring is usually added. Fruits in which the effect of the preservative is negligible are those which are yellow and, to a lesser extent, blue and purple. The powdered colouring as purchased is dissolved in hot water before incorporating it into the mix. A simple check should be made to confirm that the solid dissolves completely without leaving any sediment.

Most of the tests on dyes are performed to confirm that the material received corresponds with that supplied previously, so that the final product has a consistent colour. This applies to the detection of the components present in solution by paper chromatography (see *Method 3.4a*). It is important to check the consistency of the tinctorial power as made up and after being submitted to the effect of heat and light.

METHOD 9.6. ASSESSMENT OF THE TINCTORIAL POWER OF COLOURINGS

Make up a solution of the solid colouring by completely dissolving 1 g in 500 ml of water (solution A). Dilute A to give a concentration in the range 0·001–0·003% (solution B) and find the wavelength of maximum absorption, preferably by using a recording spectrophotometer. Adjust the concentration ($X\%$) so that the absorbance is between 0·4 and 0·65 (solution C). A concentration of 0·0025% at 515 nm was found to be suitable in the example given in *Figure 9.1.* Then measure the absorbance at the peak wavelength of

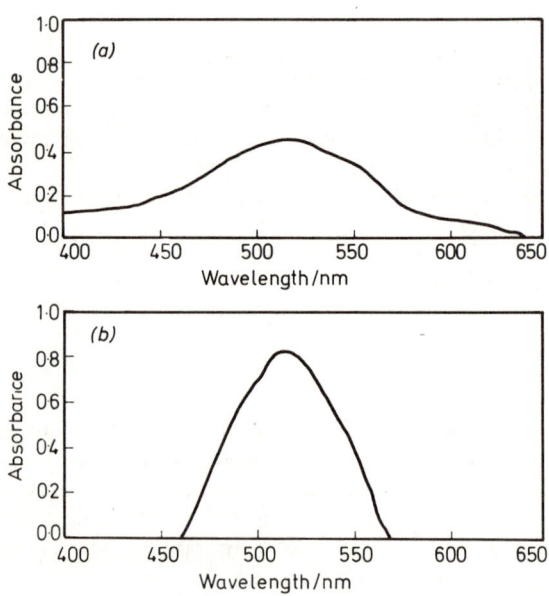

Figure 9.1. Typical absorbance curves of (a) a 0·0025% solution and (b) a 0·001% solution of a solid colouring for blackcurrant jam showing maximum peaks at 515 nm

solutions containing 0·8%, 0·9%, 1·0%, 1·1% and 1·2$X\%$ of the solid colouring. These absorbances are plotted against concentration expressed as a percentage of the standard colouring present. Thus, in the example given in *Figure 9.2*, 100% on the graph represents the suitable reference concentration of 0·0025% of the colouring in solution, and 110% represents 1·1 × 0·0025% (= 0·00275%), and so on. *Figure 9.2* can then be considered as the standard reference graph made up from the first delivery of the colouring against which all subsequent deliveries can be compared. When later deliveries

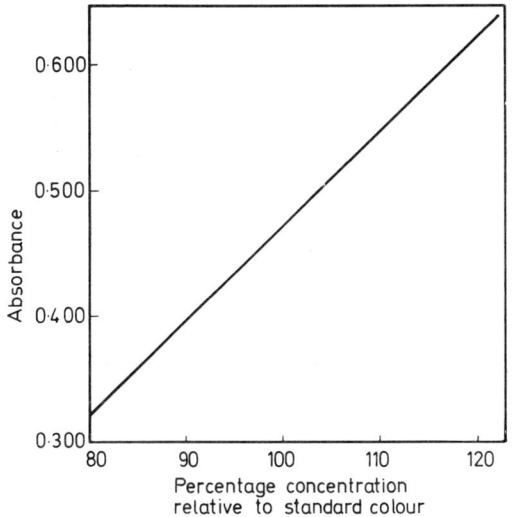

Figure 9.2. The standard graph of the tinctorial power of a blackcurrant colouring for jam manufacture. The value of 100% on the abscissa represents an actual solid dye concentration of 0·0025%. Subsequent deliveries are made up at the standard concentration of 0·0025% and the result is expressed as the percentage concentration relative to the earlier standard

are received, therefore, the colouring is made up at $X\%$ concentration and the absorbance is again measured at the peak wavelength. In the case of the colouring represented in *Figure 9.2*, if a 0·0025% solution of the colouring gives an absorbance of 0·40, this would be approximately equal to 91% of the standard. To obtain the same tinctorial power as that of the standard, it would be necessary to use 100/91 of the normal recipe amount.

It is advisable to retain some of the solid colouring used for the preparation of the standard graph for reference purposes.

Stability to heat and acid To 100 ml of the 0·2% solution A of the colouring add 1 g of citric acid, bring the mixture to the boil and boil it gently for 10 min. Cool the solution, make up the volume to the original volume, mix and measure the absorbance of a suitable dilution at the peak wavelength. Dilute A in a similar manner and compare the absorbance with that of the boiled sample.

Stability to light Expose 200 ml of the 0·2% solution A of the colouring to a strong light for several weeks. Measure the absorbence of suitable dilution of A after selected periods, say every 2 weeks. At the same time, measure the absorbance of a further volume that has been stored in a dark cupboard.

247

PROCESS AND PRODUCT CONTROL

In formulating the recipe, it is necessary to calculate the amount of soluble solids present in each ingredient, e.g., solid sugar 100%, sugar syrup 66%, pectin concentrate 10%, fruit pulp 12%. Then, from the proportions used together with water and other ingredients, it is possible to calculate the changes that occur in composition as water is lost during the boiling. The boiling is continued until the boiling temperature has increased to a final value of at least 220 °F, to correspond with a soluble solids content of at least 66·5%. A common final boiling temperature is 222 °F, which is equivalent to the boiling point of a solution containing about 70% of sugar. Lower final boiling temperatures apply, of course, if a vacuum boiling process is used. Apart from the tests mentioned below, the set of the gel is very important.

SOLUBLE SOLIDS

For legal purposes, the term 'soluble solids' means the percentage of solids ascertained with a refractometer at 20 °C, no correction being made for insoluble solids (see *Tables 2.4* and *2.4A*, pp. 59, 60). If the 'true' value is required, it is necessary to allow for the effect of other soluble ingredients (see *Table 9.1*). Such corrections are not used in practice in routine control or for legal purposes.

Table 9.1. CORRECTIONS TO BE APPLIED TO REFRACTOMETRIC SOLIDS READINGS WHEN OTHER INGREDIENTS ARE PRESENT[12] (By courtesy of the SAC).

Substance	Amount in sample, %	Correction to be applied to the total soluble solids from the direct refractometric readings, %
Invert sugar	20	+0·45
	40	+0·90
	60	+1·35
Glucose solids	20	−0·25
	40	−0·5
Citric acid	1	+0·1
	2	+0·2
	3	+0·3

Legally, jam in hermetically sealed containers must contain at least 65% of soluble solids (in 'open pack' jam, at least 68·5%) or else yeasts and moulds are liable to develop. Vacuum boiling pans are often fitted with direct-reading refractometers, so that the soluble solids content can be checked at any time during the process.

REDUCING SUGARS

During the boiling, sucrose is partially converted to invert sugar. The manufacturers aim to attain about 25–40 % of reducing sugars (as invert sugar) in the product. Rauch[7] states that the level should be kept preferably between 28 and 32 %. If there is less than about 20 %, crystals of sucrose are liable to form in the stored product; if over 40 %, a mass of invert sugar rather like granulated honey may separate on storage. The formation of invert sugar can be accelerated by one of the following means:

(a) adding more acid;

(b) adding more water in the original mix, thus increasing the boiling time;

(c) replacing some of the sucrose by invert sugar (this is often necessary in vacuum boiling owing to the lower temperature used).

The reverse applies if it is required to reduce the proportion of invert sugar formed.

In the laboratory, 15 g of sample is dissolved in cold water and the solution is transferred to a 1-litre volumetric flask. After making up the volume to the mark with water, mixing and filtering (if necessary), the solution is poured into a 50-ml burette and titrated against 10 ml of Fehling's solution by Lane and Eynon's method (*Method 2.8a*).

pH VALUE

The pH of jam (as determined on a 20 % w/v solution) should lie between 3·0 and 3·4. At low pH values (say 2·8 or below), jam is liable to 'weep', i.e., liquid tends to separate from the set gel in the jar. On the other hand, at pH 3·5 or above, it is difficult to attain a firm gel. The pH of fruits varies from about 2·5 to 4·5 (*Table 9.2*). Too high a pH as indicated in the trial boil suggests the need for the addition of citric or tartaric acid. If the pH is too low, 'buffer salts' such as sodium citrate or bicarbonate can be used.

Table 9.2. TYPICAL pH VALUES AND INSOLUBLE SOLIDS CONTENT FOR FRUITS

Fruit	pH value	Potassium, mg $K_2O/100$ g (mean figures)	Insoluble solids, % (mean figures)
Apple	3·3	158	2·03
Blackberry	3·5	185	8·0
Blackcurrant	3·1	—	5·8
Cherries	3·1–4·4	220	1·81
Damson	3·0	308	1·93
Gooseberry	3·0	200	2·41
Grape	3·8	264	—
Grapefruit	3·2	—	—
Greengage	3·2	—	1·51
Lemon	2·5	—	—
Loganberry	3·1	205	6·4
Orange	2·7	—	—
Peach	3·5	231	—
Pear	4·0	—	—
Plum	3·0–3·6	235	1·2
Quince	3·2	205	—
Raspberry	3·2	195	5·6
Redcurrant	3·0	—	5·9
Rhubarb	3·2	—	—
Strawberry	3·6	195	2·14
Tomato	4·2	—	—

SULPHUR DIOXIDE

Although up to 3000 ppm of SO_2 may have been present in the fruit pulp, most of this should be removed during the boiling and the legal maximum for the residual concentration in the final jam is 100 ppm. The loss tends to be lower if vacuum boiling is practised than when an open pan is used. The SO_2 content can be most readily estimated by the iodine titration method (*Method 3.1b*).

FRUIT CONTENT

Legal requirements for jam require the final product to contain at least 25–40% of fruit, e.g., 100 lb of strawberry jam must have been prepared with at least 38 lb of raw strawberry. Analytically, the fruit content can be assessed by determining a jam component that is preferably absent in all the other ingredients. The potassium content has often been recommended[13–15], but the element may be present in other ingredients, e.g., potassium metabisulphite is

250

sometimes used as a preservative for the fruit. In factory control, it is much easier to make allowances for such interference or merely to compare the total potassium contents for all jams made to the same recipe. Although unsuitable for jelly jam and marmalade (with its variable peel content), the mean insoluble solids content is probably more generally useful for assessing the fruit content. As with all the values suggested, however, the insoluble solids contents of fruits show wide natural variations from averages.

METHOD 9.7. ESTIMATION OF THE INSOLUBLE SOLIDS CONTENT OF JAM

Commence by washing a 15-cm Whatman No. 4 filter-paper with water and place it in a 100 °C oven. This is required later—the paper is subsequently dried for 2 h, cooled in a desiccator, weighed in a large, dried, stoppered weighing bottle and used for the filtration below.

Weigh out 25 g of the blended jam into a 400-ml beaker, add about 200 ml of hot water and boil the mixture gently for 15 min. Mix with a rod and filter the mixture down the rod into the prepared, weighed filter-paper. After collecting the insoluble matter in the paper and passing all the filtrate through, wash the paper *very thoroughly* with what may seem to be an excess of hot water. Then dry the paper overnight in the 100 °C oven and weigh as before in the same dried, stoppered bottle after cooling in a desiccator. Calculate the insoluble solids content as a percentage of the jam sample. Then assess the approximate fruit content of the jam from the following equation:

$$\text{Calculated fruit } (\%) = \frac{\% \text{ Insoluble solids in jam}}{\text{Mean } \% \text{ insoluble solids in fruit}} \times 100$$

Typical mean percentages for the insoluble solids and potassium contents of various fruits are given in *Table 9.2*. As the natural fruits are known to show wide variations from such mean values (for example, the insoluble solids content of raspberry may vary from 1·4 to 9·2 %), the use of such values can give only an approximate indication of the amount of raw material actually used (see Money[13]).

CANNED FRUITS

The sequences involved in the manufacture of canned fruit are as follows:

Preparation of the fruit (sorting, grading, peeling, trimming, washing)

Blanching

Filling of fruit into cans

Syruping (filling of syrup into cans up to a definite height)

Exhausting

Lidding and sealing (seaming)

Heat processing at 212 °F

Cooling and labelling.

Although under-ripe fruit shrivels or toughens during processing, over-ripe fruit is also undesirable. The latter breaks down during the heating whereas a good final product should still have a firm flesh, but be ready to eat. In general, the fruit used should, as far as possible, be of uniform colour, maturity and size. The weight of fruit filled into the can should comply with the minimum weight stated in LAJAC Code of Practice No. 4[16]. The blanching or pre-cooking causes the fruit to soften and shrink and allows the can to be well filled. The operation also partially sterilises the fruit and drives out occluded air, which otherwise would encourage subsequent corrosion of the can. Four strengths of syrup are included in the LAJAC Code, viz. (all % w/w = °Brix), Light Syrup 10–15%, Syrup 15–30%, Heavy Syrup 20–40% and Extra Heavy Syrup 40–50%, according to the fruit. The syrup is prepared hot and the strength is checked with a Brix hydrometer or refractometer after correcting to give the reading corresponding to 20 °C (*Tables 2.4, 2.4A, 9.3* and *9.4*). The cans are filled to within $\frac{1}{4}$–$\frac{5}{16}$ in. of the top of the can before rolling on the lid. This gives a final headspace of approximately $\frac{1}{8}$–$\frac{3}{16}$ in., the remainder being taken up by the countersink of the lid. Hydrogen swells ('springers') are more likely to arise in overfilled cans that have too small a headspace. Exhausting involves heating the open, filled cans in water at about 185 °F.

The cans emerge at about 160 °F so that steam largely replaces air in the headspace before seaming. In view of the low pH of fruits, processing at 212 °F is sufficient to render the pack commercially sterile. The cans are then cooled with cold water. Insufficient cooling causes 'stack-burning', which encourages corrosion of the can and affects the colour of the product.

RAW MATERIALS

Materials such as the sugar (p. 237) and any added colouring (p. 245) can be examined as described earlier in this Chapter.

Table 9.3. SUGAR SYRUP—BRIX AND HYDROMETER SCALES AND BOILING POINTS

Degrees Brix (% sugar by weight) at 20°C (68°F)	Specific gravity at 20°C (68°F)	Degrees Baumé at 20°C (68°F)	Degrees Twaddell at 20°C (68°F)	Wt. of sugar added to 1 Imp. gal. of water		Wt. of sugar in 1 Imp. gal. of syrup		Boiling point of syrup	
				lb	oz	lb	oz	°C	°F
10	1·040	5·6	8·0	1	2	1	½	100·1	212·2
12	1·048	6·7	9·6	1	6	1	4	—	—
14	1·057	7·8	11·4	1	10	1	7½		
16	1·065	8·9	13·0	1	14	1	11		
18	1·074	10·0	14·8	2	3	1	14½		
20	1·083	11·1	16·6	2	8	2	2½	100·3	212·5
22	1·092	12·2	18·4	2	13	2	6		
24	1·101	13·3	20·2	3	3	2	10		
26	1·110	14·4	22·0	3	8	2	14		
28	1·119	15·5	23·8	3	14	3	2		
30	1·129	16·6	25·8	4	5	3	6		
32	1·139	17·7	27·8	4	11	3	10	100·6	213·1
34	1·148	18·7	29·6	5	3	3	14		
36	1·158	19·8	31·6	5	10	4	2½		
38	1·168	20·9	33·6	6	2	4	7		
40	1·179	22·0	35·8	6	11	4	11		
42	1·189	23·0	37·8	7	4	5	0	101·1	214·0
44	1·199	24·1	39·8	7	14	5	4½		
46	1·210	25·2	42·0	8	9	5	9		
48	1·221	26·2	44·2	9	4	5	13½		
50	1·232	27·3	46·4	10	0	6	2	101·9	215·5
52	1·243	28·3	48·6	10	14	6	7		
54	1·254	29·4	50·8	11	12	6	12		
56	1·265	30·4	53·0	12	12	7	1		
58	1·277	31·5	55·4	13	14	7	6		
60	1·288	32·5	57·6	15	1	7	11½	103·1	217·4
62	1·301	33·5	60·2	16	6	8	1		
64	1·313	34·5	62·6	17	13	8	6		
66	1·325	35·6	65·0	19	8	8	11½		
68	1·337	36·6	67·4	21	5	9	1½		
70	1·350	37·6	70·0	23	6	9	7	105·3	221·6
75	1·381	40·0	76·2					107·4	225·3
80	1·414	42·5	82·8					110·3	230·5
85	1·448	44·9	89·6					114·5	238·1
90	1·483	47·2	96·6					122·6	252·7

253

Table 9.4. TEMPERATURE CORRECTIONS FOR BRIX HYDROMETER READINGS

	Temp., °F	Degrees Brix and Correction									
		10	20	25	30	35	40	45	50	55	60
Subtract	40	0·5	0·6	0·7	0·8	0·9	0·9	0·9	0·9	0·9	1·0
correction	50	0·5	0·5	0·5	0·5	0·6	0·6	0·6	0·6	0·6	0·6
	60	0·2	0·2	0·2	0·2	0·2	0·2	0·2	0·2	0·2	0·2
Add correction	70	0·1	0·2	0·2	0·2	0·2	0·2	0·2	0·2	0·2	0·2
	80	0·5	0·6	0·6	0·6	0·6	0·6	0·6	0·6	0·6	0·6
	90	0·9	1·0	1·0	1·0	1·1	1·1	1·1	1·1	1·1	1·0
	100	1·3	1·4	1·5	1·5	1·5	1·5	1·5	1·5	1·5	1·5
	120	2·5	2·6	2·6	2·6	2·6	2·6	2·6	2·6	2·5	2·5
	140	3·8	3·8	3·8	3·8	3·8	3·8	3·7	3·7	3·6	3·6
	160	5·1	5·1	5·1	5·1	5·1	5·0	5·0	4·9	4·8	4·8
	180	6·7	6·5	6·4	6·4	6·3	6·3	6·2	6·1	6·0	5·9
	212	10·0	9·6	9·4	9·3	9·1	8·9	8·7	8·4	8·2	8·1

For temperatures below 68 °F (20 °C), subtract the correction. For temperatures above 68 °F (20 °C), add the correction.

EXAMINATION OF TRIAL BATCHES

The mean 'refractometer solids' of raw fruits varies from about 8 to 17%. Usually the sugar concentration in the added syrup is greater. During the processing, therefore, the syrup strength is reduced by the weaker fruit juice. Eventually, the total sugar in the can is evenly distributed through the whole can—both fruit and liquor (the cut-out strength). Morris[17] recommended using the following equations for calculating the initial syrup strength to give the desired final strength (or vice versa):

$$W_1 + W_2 = W_3 \tag{9.1}$$
$$xW_1 + yW_2 = zW_3 \tag{9.2}$$

where W_1 = weight of fruit in can
W_2 = weight of added syrup in can
W_3 = weight of fruit + syrup in can
x = soluble solids (%) in the fruit juice of the fruit after cooking, allowance being made for insoluble solids
y = sugar in syrup (%)
z = soluble solids in the final product (%)

More generally:

W_I = weight of fruit per unit volume of final product
W_{II} = weight of syrup per unit volume of final product

254

W_{III} = weight of a unit volume of the final syrup, i.e., the specific gravity of the final syrup

The value of x (the soluble solids in the fruit juice after cooking) should be determined on a large number of samples taken from the new season's fruit. The following are approximate values obtained for x: blackberry 13, blackcurrant 15, cherry 10, damson 11·5, greengage 12, loganberry 10·5, plum 10, raspberry 8, redcurrant 14.

As an example, suppose that it is proposed to produce blackberries with a final cut-out strength of 27%. Taking W_I, W_{II} and W_{III} on the weight per unit volume basis:

W_I = 0·59 oz

W_{III} = 1·115 oz (the sp. gr. of 27% sugar solution—from *Table 9.3*)

W_{II} = 1·115 − 0·59 = 0·525 oz

From equation (9.2) and taking x = 13 for blackberries:

$$13 \times 0·59 + y \times 0·525 = 27 \times 1·115$$

$$y = 42·8\%$$

The use of such a strength of added syrup would correspond to a product described as 'Blackberries in Heavy Syrup'. In making adjustments, a fall of 5° Brix in the added syrup causes a drop of about 1% in the drained weight and a fall of about 2·5–3·0% Brix on cut-out.

PRODUCT

The full can is first weighed (W_T g). The vacuum is then measured by gauge. One end of the can is carefully removed and the headspace measured. The contents are then poured on to a sieve that has 8 holes per square inch and the syrup is collected. After remaining on the sieve for 2 min, the drained fruit is weighed (W_D g). The can plus lid is then weighed after washing them with hot water and drying with hot air (W_C g). The net weight is calculated as $\dfrac{W_T - W_C}{28·35}$ oz. Then the probable filled weight is calculated as $\dfrac{100 W_D}{D}$, where D is the mean percentage drained weight as a percentage of the filled weight. Typical values of *D* are as follows: blackberry 81, black-currant 86, cherry 90, damson 90, gooseberry 94, greengage 93, loganberry 82, plums 86, raspberry 82 and strawberry 66 (cf., Adam[18]). These values apply to products in which the added syrup concentration is 40° Brix. Fruits packed in lighter syrups have drained weights that are 2–5% greater than those above. More generally, in addition to the syrup density and the time of draining on the sieve, the drained weight percentage varies according to the variety, condition, ripeness, etc., of the fruit, the processing and storage conditions, and so on. The calculated filled weight should be compared with the value stated in the LAJAC Code of Practice[16], which calls for the examination of 10 cans, from which the mean is to be taken.

The inner surface of the can should be examined for stains such as 'feathering' or dull-grey rough patches. In any instances of doubt, the material (fruit mixed with the normal proportion of syrup) should be examined for tin (*Method 4.9a*), for which the maximum level recommended by the FSC is 250 ppm (see below). Any imperfections in the lacquer should be examined closely. Fruit lacquers can usually be removed with carbon tetrachloride. Any solder splashes should also be noted.

The separated syrup should also be examined for pH, acidity and refractometric solids. In certain instances, the mixed contents should be examined for added colouring (p. 90), iron (p. 105) and copper (p. 103). The initial rate of dissolution of tin is rapid, but is very much slower a few days after canning when the residual oxygen in the can has been used up. Immediately after canning, the iron content of canned fruits seldom exceeds 30 ppm. In badly lacquered cans, much higher levels have been found on storage, the main attack often occurring at the seams.

PICKLES AND SAUCES

In the manufacture of pickles, the vegetables are fermented in brine containing 10% of salt. The salt concentration is later raised gradually to 15%. After removing excess of salt, the vegetables are placed in vinegar containing 5% of acetic acid. After a further period, the liquid is drained off and the material cut prior to covering with 3% acetic acid.

Sauces are manufactured from vinegar, sugar, spices, salt, fruits and vegetables and a thickening agent. They are less acidic than sauces and require heat treatment at about 180 °F. On the average, sauces contain 2·7% of acid and 25% of sugar.

INGREDIENTS

The control of the salt and sugar used as raw materials is discussed on pages 185 and 237, respectively. The treated vegetables can be examined for salt and sugar as for the final pickles (see later).

VINEGAR

The *total acidity* (Section 2.9) is determined by titrating 10 ml of sample with 0·5 N NaOH using phenolphthalein as indicator.

Total acidity ($\%$ w/v as acetic acid) = Titration × 0·3

The minimum acidity usually accepted by the courts is 4·0$\%$[19]. Vinegars of reasonable quality usually have an acidity of about 5$\%$.

A simple method for differentiating between brewed vinegars, such as malt vinegar, and non-brewed products is to apply the formol titration in the following way.

METHOD 9.8. APPLICATION OF FORMOL TITRATION TO VINEGAR

Pipette 10 ml of sample into a porcelain dish, add phenolphthalein solution, *nearly* neutralise the solution with 0·5 N NaOH and then render it *exactly* neutral with 0·05 N NaOH. Exactly neutralise some formalin with 0·05 N NaOH and add 5 ml to the neutralised sample. Stir the mixture, allow it to stand for 5 min, then titrate the acidity produced with 0·05 N NaOH. Non-brewed samples and distilled vinegars give a negligible formol titration. If, however, the second titration is appreciable (e.g., more than 0·7 ml of 0·05 N NaOH), a brewed vinegar is indicated.

More accurately, it is possible to differentiate between the various types of vinegars by making determinations on the distillate[20], e.g., oxidation value[21], ester value[21], iodine value[21] and the alkaline oxidation value[22], which is determined as follows.

METHOD 9.9. DETERMINATION OF ALKALINE OXIDATION VALUE OF VINEGAR[22]

Distil 60 ml of sample from a 400-ml flask, fitted with a tap funnel, into a 100-ml graduated cylinder. When 45 ml of distillate has been collected, add 15 ml of water to the flask through the tap funnel and continue the distillation until the total volume of distillate is exactly

60 ml. Obtain two 300-ml glass-stoppered bottles, A and B. To A only, add by pipette 2·0 ml of mixed distillate. Then treat both A and B as follows. Add about 100 ml of water and 10 ml of 10% NaOH, and then by pipette 10 ml of 0·1 N potassium permanganate solution. After allowing the solutions to stand for 30 min, acidify them with 10 ml of dilute sulphuric acid (concentrated acid diluted 1 + 3 with water), mix and add 0·5 g of KI. Titrate the liberated iodine with 0·02 N sodium thiosulphate using starch as indicator near the end-point. If the sample and blank titrations are A and B, respectively:

$$\text{Alkaline oxidation value} = 8(B - A)$$

Malt vinegars usually give values from 50 to 260, but non-brewed samples give values below 15 (cf., White[23]).

TOMATO PURÉE

Tomato purée is made by concentrating the strained liquid pulp of good quality tomatoes. The raw tomato (5% total solids) is converted to 'double concentrate' (28–30% total solids) and 'triple concentrate' (36–40% total solids). The quality is assessed from the total solids (see below), sugars before and after inversion (by Lane and Eynon's method, *Method 2.8a*), salt and acidity (see below), pH of a 20% solution, copper (*Method 4.4*), organoleptic quality including colour (see p. 260) and the Howard Moult Count[24–26]. Good, even sampling is especially important with tomato purée.

METHOD 9.10. ESTIMATION OF TOTAL SOLIDS IN TOMATO PURÉE WITH A REFRACTOMETER

If the total solids content does not exceed 20%, squeeze the sample through a double thickness of muslin on to the prism of an Abbé refractometer. With more concentrated products, dilute 10·0 ± 0·05 g of sample with 20·0 ± 0·05 g of water, mix well and filter (rejecting the first few drops) through a small filter-paper on to the refractometer. Obtain the total solids by reference to *Table 9.5*.

METHOD 9.11. ESTIMATION OF SALT AND ACIDITY IN TOMATO PURÉE

Weigh out 10 g of purée into a 250-ml beaker. Add about 90 ml of water, mix with a rod, warm the mixture slightly and transfer it

Table 9.5. TOTAL SOLIDS IN TOMATO PURÉE ESTIMATED FROM REFRACTOMETER READINGS AT 20 °C (AFTER SIPPLE[27])

Filtrate from whole purée		Filtrate from purée diluted 1:2 at 25 °C	
Refractive index	Total solids, %	Refractive index	Total solids, %
1·34520	9·0	1·34204	20·5
1·34588	9·5	1·34227	21·0
1·34656	10·0	1·34250	21·5
1·34724	10·5	1·34273	22·0
1·34794	11·0	1·34296	22·5
1·34862	11·5	1·34319	23·0
1·34930	12·0	1·34342	23·5
1·35000	12·5	1·34365	24·0
1·35069	13·0	1·34388	24·5
1·35139	13·5	1·34411	25·0
1·35209	14·0	1·34434	25·5
1·35280	14·5	1·34457	26·0
1·35352	15·0	1·34480	26·5
1·35425	15·5	1·34503	27·0
1·35497	16·0	1·34526	27·5
1·35571	16·5	1·34550	28·0
1·35644	17·0	1·34573	28·5
1·35717	17·5	1·34596	29·0
1·35792	18·0	1·34620	29·5
1·35867	18·5	1·34643	30·0
1·35942	19·0	1·34667	30·5
1·36020	19·5	1·34691	31·0
1·36097	20·0	1·34714	31·5
		1·34738	32·0
		1·34762	32·5
		1·34786	33·0
		1·34810	33·5
		1·34834	34·0
		1·34858	34·5
		1·34881	35·0

carefully to a 500-ml volumetric flask. Cool, make up the volume to 500 ml, mix and filter (filtrate A).

To 50 ml of A (= 1 g of sample) add potassium chromate solution and a piece of marble and titrate with 0·05 N silver nitrate solution.

$$1 \text{ ml of } 0\cdot05 \text{ N AgNO}_3 \equiv 0\cdot00292 \text{ g of NaCl}$$

Titrate a further 50 ml of A with 0·1 N NaOH using phenol-phthalein as indicator.

$$1 \text{ ml of } 0\cdot1 \text{ N NaOH} \equiv 0\cdot0070 \text{ g of citric acid}$$

The titratable acidity of purée of reasonable quality should not exceed 11 % as citric acid when expressed on the salt-free solids.

259

METHOD 9.12. ESTIMATION OF COLOUR IN TOMATO PURÉE USING THE LOVIBOND TINTOMETER

Place the Lovibond Tintometer on its legs in the upright position (with the eyepiece horizontal) and match the colour of the sample after pouring it into one of the flat porcelain trays. A full rich red sample normally gives no more than about 5 Units of yellow + equal red. Some organisations prefer to use the Gardner Colour Difference Method (Wentworth Instruments Ltd., Datchet, Bucks.).

PRODUCTS

PICKLES

After measuring the total weight (preparatory to obtaining the net weight later), the contents should be examined for general appearance of the pack. Then, after checking the vacuum (for pasteurised or vacuum-sealed products), the lid should be removed and examined for corrosion. For appropriate products, such as pickled onions, the drained weight is assessed as for canned fruit (p. 255). In other instances, e.g., with products in thick sauce, the sauce can be washed off with warm water, although this procedure, of course, affects subsequent determinations. From such weights, the ratio of the drained weight of vegetables to liquor should be calculated. Also, the individual vegetables may sometimes be separated and weighed and the proportions of each calculated. Other estimations that may be required are those for trace elements and preservatives (pp. 78–118, including *Method 3.2a*, p. 85).

In sampling, the material should be rendered homogeneous in a mechanical blender. Products in thick liquor should be squeezed through a double thickness of muslin on to the prism of an Abbé refractometer and the refractometric soluble solids estimated as sucrose in the usual way. If the amount is significant, the sugars should be further estimated before and after inversion (after clearing) by Lane and Eynon's method, *Method 2.8a*. Then, from the approximate water content (see *Method 9.14*), the following ratio can be calculated:

$$\text{Sugars in aqueous phase } (\%) = \frac{\% \text{ Total sugars}}{\% \text{ Total sugars} + \% \text{ Water}} \times 100$$

The salt and volatile acidity should also be expressed on the aqueous phase as indicated in the following methods.

METHOD 9.13. ESTIMATION OF SALT IN PICKLES

Weigh 10 g of a representative sample into a porcelain basin, add water and, following a preliminary titration, 0·1 ml less of 0·5 N NaOH than the volume required to just neutralise the solution. Then add 2 ml of 5% potassium chromate and titrate to the first sign of an orange colour with 0·1 N silver nitrate.

$$1 \text{ ml of } 0 \cdot 1 \text{ N AgNO}_3 \equiv 0 \cdot 00585 \text{ g of NaCl}$$

From the approximate water content (see *Method 9.14*), calculate the following:

$$\text{Salt in aqueous phase } (\%) = \frac{\% \text{ NaCl}}{\% \text{ NaCl} + \% \text{ Water}} \times 100$$

METHOD 9.14. ESTIMATION OF THE TOTAL VOLATILES AND THE TOTAL AND FIXED ACIDITY OF PICKLES

Total volatiles (water + acetic acid, etc.)

Weigh out 5 g of sample into a prepared metal dish dried previously with sand and containing a rod. Add water, stir to mix and dry the mixture, first on a water-bath and then for 3 h in an oven at 100 °C. Cool in a desiccator, weigh and calculate the percentage weight lost, i.e., the total volatile content.

Total acidity

Weigh 10 g of sample into a porcelain dish, add water, stir and titrate the solution with 0·5 N NaOH using phenolphthalein as indicator. Calculate the acidity as acetic acid.

$$1 \text{ ml of } 0 \cdot 5 \text{ N NaOH} \equiv 0 \cdot 030 \text{ g of acetic acid}$$

Fixed acidity

Weigh 10 g of sample into a porcelain dish, add 30 ml of water, stir and evaporate the mixture on a water-bath. Add a further 30 ml of water, stir the residue well into the liquid and evaporate the mixture again. Repeat the evaporation at least twice more. Then finally re-stir the residue with more water and titrate the solution as for the total acidity to give the fixed acidity as acetic acid.

$$\text{Volatile acidity (\%)} = \text{\% Total acidity} - \text{\% Fixed acidity}$$
(as acetic acid)   (as acetic acid)   (as acetic acid)

$$\text{Volatile acidity in aqueous phase (\%)} = \frac{\text{\% Volatile acidity}}{\text{\% Total volatiles}} \times 100$$
(equilibrium acidity)

For good keeping qualities, the equilibrium acidity should be at least 3·5%. Lower values may be appropriate, however, with products that contain significant amounts of sugar and/or salt or if the product has been heat-processed[28].

SAUCES

Apart from examination for viscosity and, in the case of tomato ketchup, the estimation of copper, tomato solids (from the potassium value) and the Howard Mould Count[24], the determinations described for pickles are in general also applicable to sauces.

CITRUS FRUIT JUICES

Juices prepared from citrus fruits (orange, lemon, grapefruit and lime) are prepared by expression (reaming), straining and flash pasteurisation. The heat treatment destroys the pectic enzymes, which otherwise would tend to make the juice clarify (separate) on storage instead of retaining its cloudiness. A large number of analytical methods are available for checking the quality of juices, e.g., specific gravity, refractive index, total solids, alcohol, total and volatile acidity, pH, sugars, ash, chloride, nitrogen (and formol titration), volatile oil, pectin, enzymic activity, hydroxymethylfurfural, phosphorus, potassium, vitamin C, sulphur dioxide and benzoic acid. Some typical values obtained with citrus juices are given in *Table 9.6*. Special points relating to some of these determinations are given below.

Table 9.6. TYPICAL AVERAGE VALUES OBTAINED FROM FRESH CITRUS FRUIT JUICES

Juice	Specific gravity	Total solids content, % w/w	Total acidity (as citric acid), % w/w	Ash, % w/w	Vitamin C, mg/100 ml
Grapefruit	1·040	10·4	1·6	0·47	41
Lemon	1·035	10·0	4·9	0·37	46
Lime	1·035	9·3	7·5	0·38	25
Orange	1·042	10·8	1·4	0·40	48

SPECIFIC GRAVITY AND REFRACTIVE INDEX

The sugars can be most accurately determined (before and after inversion) after clearing with zinc ferrocyanide by using the Lane and Eynon method (*Method 2.8a*, p. 62). For routine purposes, the total sugars can be estimated indirectly from the specific gravity (Brix hydrometer) or refractive index (*Tables 2.4* and *9.3*). *Table 9.7* gives values that can be used for assessing the degree of concentration of concentrates.

ACIDITY

After removing carbon dioxide by boiling and cooling the fruit juice, the titratable acidity can be determined by using 0.1 N NaOH, with phenolphthalein as indicator. The acidity is usually calculated as citric acid.

$$1 \text{ ml of } 0.1 \text{ N NaOH} \equiv 0.0070 \text{ g of citric acid}$$

MATURITY RATIO

The flavour of fruit juices can best be considered from the maturity ratio, i.e., the ratio of the total solids (or Brix hydrometer reading) to the acidity (as citric acid). This ratio assesses the balance of flavour between the sweetness and the acidity and increases during ripening and changes during the season for fruit at the same stage of maturity.

Grading systems for fruits and juices based on limits for sugar (Brix readings) and acidity are applied by many exporting countries.

ENZYMIC ACTIVITY

The following method for pectolytic activity has been recommended by Dickinson and Goose[29].

METHOD 9.15. ASSESSMENT OF RESIDUAL PECTOLYTIC ACTIVITY OF FRUIT JUICES[29]

Citrus pectin solution Dissolve 5 g of citrus pectin and 5.85 g of NaCl in water and dilute the solution to 500 ml.

Table 9.7. DEGREE OF CONCENTRATION OF CITRUS JUICES FROM THE SPECIFIC GRAVITY (20°/20 °C) AND REFRACTIVE INDEX AT 20 °C

Concentration	Lemon		Orange		Grapefruit	
	Specific gravity	Refractive index	Specific gravity	Refractive index	Specific gravity	Refractive index
$\times 3$	1·111–1·120	1·3681–1·3710	1·139–1·150	1·3816–1·3860	1·121–1·130	1·3756–1·3790
$3\frac{1}{4}$	1·121–1·128	1·3711–1·3740	1·151–1·163	1·3861–1·3905	1·131–1·143	1·3791–1·3835
$3\frac{1}{2}$	1·129–1·135	1·3741–1·3770	1·164–1·175	1·3906–1·3950	1·144–1·155	1·3836–1·3880
$3\frac{3}{4}$	1·136–1·143	1·3771–1·3795	1·176–1·188	1·3951–1·3995	1·156–1·168	1·3881–1·3925
4	1·144–1·150	1·3796–1·3820	1·189–1·200	1·3996–1·4040	1·169–1·180	1·3926–1·3970
$4\frac{1}{4}$	1·151–1·160	1·3821–1·3850	1·201–1·213	1·4041–1·4085	1·181–1·190	1·3971–1·4010
$4\frac{1}{2}$	1·161–1·170	1·3851–1·3880	1·214–1·225	1·4086–1·4130	1·191–1·200	1·4011–1·4050
$4\frac{3}{4}$	1·171–1·180	1·3881–1·3910	1·226–1·238	1·4131–1·4180	1·201–1·210	1·4051–1·4085
5	1·181–1·190	1·3911–1·3940	1·239–1·250	1·4181–1·4230	1·211–1·220	1·4086–1·4120
$5\frac{1}{4}$	1·191–1·200	1·3941–1·3975	1·251–1·263	1·4231–1·4280	1·221–1·230	1·4121–1·4160
$5\frac{1}{2}$	1·201–1·210	1·3976–1·4010	1·264–1·275	1·4281–1·4330	1·231–1·240	1·4161–1·4200
$5\frac{3}{4}$	1·211–1·220	1·4011–1·4040	1·276–1·288	1·4331–1·4380	1·241–1·250	1·4201–1·4235
6	1·221–1·230	1·4041–1·4070	1·289–1·300	1·4381–1·4430	1·251–1·260	1·4236–1·4270
$6\frac{1}{4}$	1·231–1·238	1·4071–1·4095	1·301–1·313	1·4431–1·4475	1·261–1·270	1·4271–1·4310
$6\frac{1}{2}$	1·239–1·245	1·4096–1·4120	1·314–1·325	1·4476–1·4520	1·271–1·280	1·4311–1·4350
$6\frac{3}{4}$	1·246–1·253	1·4121–1·4148	1·326–1·338	1·4521–1·4565	1·281–1·290	1·4351–1·4390
7	1·254–1·260	1·4149–1·4175	1·339–1·350	1·4566–1·4610	1·291–1·300	1·4391–1·4430

Procedure

Heat 50 ml of citrus pectin solution in a beaker, maintain the temperature at 95 °C for 1 min, then cool the solution rapidly to 30 °C. Add 2 ml of juice (with concentrates, dilute them first to their original strength), mix and maintain at 30 °C throughout the whole of the subsequent operations. Using a pH meter, bring the solution to pH 7·0 by the addition of 0·02 N NaOH from a burette. Note the burette reading and continue the addition of 0·02 N NaOH over a period of exactly 30 min in order to maintain the pH at 7·0 (V ml). Calculate the activity of the juice in terms of milligrams of methoxyl split off by 1 ml of enzyme solution:

$$\text{Activity} = \frac{V \times 0·02 \times 31}{x}$$

where x = volume of juice taken = 2 ml for straight juice. Note that V is the volume of 0·02 N NaOH required to maintain a pH of 7 subsequent to neutralisation.

The pectic enzymes should be destroyed during pasteurisation. If the heat treatment is inadequate, the stored material (more particularly in squashes prepared from it) is liable to separate on storage and to lose the desirable cloudiness. A heat-treated juice has an elevated hydroxymethylfurfural content compared with an unheated juice (*Method 9.16*). '

HYDROXYMETHYLFURFURAL

Hydroxymethylfurfural is an aldehyde formed when hexoses are heated. Its determination therefore gives an indication of whether or not a fruit juice or concentrate has been heat-treated.

METHOD 9.16. DETERMINATION OF HYDROXYMETHYLFURFURAL IN FRUIT JUICES[30]

Principle of method

Hydroxymethylfurfural (HMF) reacts with barbituric acid and *p*-toluidine to give a red colour, the optical density of which can be measured at 550 nm. Sulphites interfere but, if present, the aldehyde can be set free under weakly alkaline conditions and the sulphurous acid is oxidised with iodine.

Special reagents

Barbituric acid Dissolve 500 mg of barbituric acid (dried at 105 °C) by warming it with water in a 100-ml volumetric flask on a water-bath, and make up the volume to the mark with water. The solution should be freshly prepared.

p-Toluidine Dissolve 10 g of *p*-toluidine in 50 ml of isopropanol in a 100-ml volumetric flask, add 10 ml of glacial acetic acid and make up the volume to the mark with isopropanol. The solution should be freshly prepared.

Procedure

Make 25-ml of fruit juice or the equivalent volume of concentrate *weakly* alkaline (pH 7–8) with solid sodium bicarbonate in a 50-ml volumetric flask and add a few drops of isopropanol to reduce foaming. Add 4 ml of 0·5% starch solution then add dropwise 0·1 N iodine until a blue colour persists for 15 s. Add 1 ml of each of the zinc ferrocyanide clearing agents (p. 60), mixing between each addition, make up the volume to the mark with water, mix and filter the solution. Pipette 2 ml of filtrate into each of two small flasks, A and B, fitted with glass stoppers. Add by pipette 5 ml of *p*-toluidine and 1 ml of water to B, mix and measure the optical density of the solution within 3 min at 550 nm (blank). To A, add 5 ml of *p*-toluidine and 1 ml of barbituric acid and measure the optical density in the same way (sample). Subtract the blank reading from the sample reading to obtain the HMF content in the sample. High quality juices give little or no colour, equivalent to no more than a few milligrams of HMF per litre. Higher amounts indicate that the juice has been heat-treated and should be estimated by comparison with a reference curve prepared from pure HMF (linear up to 50 mg of HMF per litre). The HMF can also be estimated in an ether extract at 284 nm[31].

METHOD 9.17. DETERMINATION OF VITAMIN C IN FRUIT JUICES

Principle of method

The juice is diluted with metaphosphoric acid, which inactivates ascorbic oxidase and the vitamin C is determined by its reducing action on the blue dyestuff 2,6-dichlorophenolindophenol. The addition of acetone prevents interference by sulphur dioxide, due to formation of the acetone–bisulphite complex.

Reagents

Metaphosphoric acid solution Prepare a 20% aqueous solution.
Standard ascorbic acid solution Dissolve 0·0500 g of pure ascorbic acid in 60 ml of cold 20% metaphosphoric acid in a 250-ml volumetric flask and dilute to the mark with water.

$$1 \text{ ml of solution} \equiv 0.2 \text{ mg of vitamin C}$$

Indophenol dye solution Dissolve 0·05 g of 2,6-dichlorophenolindophenol in cold water and dilute the solution to 100 ml. Filter the solution and standardise it by titrating it against 10 ml of standard ascorbic acid solution until a faint pink colour persists for 15 s. Express the concentration in terms of milligrams of vitamin C equivalent to 1 ml of dye solution.

Procedure

Pipette 50 ml of juice (or 10 ml or 10 g of concentrated juice) into a 100-ml volumetric flask, add 25 ml of 20% metaphosphoric acid and dilute to the mark with water. Mix, pipette 10 ml of the solution into a small conical flask, add 2·5 ml acetone and titrate with the indophenol dye solution until a faint pink colour persists for 15 s. Calculate the vitamin C content of the sample as milligrams per 100 ml or 100 g, allowing if necessary for the density of the original material.

$$1 \text{ International Unit of vitamin C} = 50 \ \mu g$$

If the juice is highly coloured, the vitamin C can be titrated potentiometrically[32]. Special methods for blackcurrant syrup have been described in the BPC and by Kum-Tutt and Leong[33].

PRESERVATIVES

Benzoic acid can be directly extracted from the acidified sample with ether or chloroform (p. 85). Sulphur dioxide should preferably be determined by one of the distillation procedures described in Chapter 3. For more rapid results, it is possible to use direct titration methods. In the following method, any bound sulphur dioxide present is first released by contact with alkali.

METHOD 9.18. ESTIMATION OF TOTAL SULPHUR DIOXIDE IN FRUIT JUICES BY DIRECT TITRATION

Add 50 ml of juice to 25 ml of 1 N NaOH and allow the mixture to stand for 10 min. Add 10 ml of dilute sulphuric acid $(1+3)$ and starch solution and titrate with 0·05 N iodine (cf., *Method 3.1c*, p. 84).

$$1 \text{ ml of } 0·05 \text{ N iodine} \equiv 0·0016 \text{ g of SO}_2$$

$$\text{Sulphur dioxide (ppm)} = \frac{\text{Titration} \times 0·0016 \times 10^6}{50}$$

ADULTERATION AND SOPHISTICATION

Various problems associated with the detection of the adulteration of juices have been discussed by the author[14]. Genuineness is most commonly confirmed from the phosphate and potassium contents. Such values have in turn been frequently used for assessing the fruit content of squashes.

REFERENCES

1. ICUMSA, *Int. Sug. J.*, **72**, No. 860, 259 (1970)
2. KNIGHT, J. and ALLEN, C. H., *Int. Sug. J.*, **62**, 344 (1960)
3. HIBBERT, D. and PHILLIPSON, R. T., *Int. Sug. J.*, **68**, 39 (1966)
4. CARRUTHERS, A., HEANEY, R. K. and OLDFIELD, J. F. T., *Int. Sug. J.*, **67**, 364 (1965)
5. STEIGMANN, A., *J. Soc. chem. Ind., Lond.*, **61**, 18 (1942)
6. WEST, P. W. and GAEKE, G. C., *Analyt. Chem.*, **28**, 1816 (1956)
7. RAUCH, G. H., *Jam Manufacture*, 2nd edn, Leonard Hill, London (1965)
8. CHATT, E. M. and HINTON, C. L., *Fd Trade Rev.*, **25**, No. 10, 3 (1955)
9. BFMIRA, *Analyst, Lond.*, **76**, 536 (1951)
10. OLLIVER, M., WADE, P. and DENT, K. P., *J. Sci. Fd Agric.*, **8**, 188 (1957)
11. MOLYNEUX, F., *Proc. Biochem.*, **6**, No. 5, 17 (1971)
12. MACARA, T., *Analyst, Lond.*, **56**, 391 (1931)
13. MONEY, R. W., *J. Sci. Fd Agric.*, **15**, 594 (1964)
14. PEARSON, D., *The Chemical Analysis of Foods*, 6th edn, Churchill, London (1970)
15. *Food Standards Committee Report on Jams and Other Preserves*, H.M.S.O., London (1968)
16. LOCAL AUTHORITIES' JOINT ADVISORY COMMITTEE ON FOOD STANDARDS, *J. Assoc. Publ. Analysts*, **3**, 106 (1965)
17. MORRIS, T. N., *Principles of Fruit Preservation*, Chapman and Hall, London (1951)

18. ADAM, W. B., *J. Assoc. Publ. Analysts*, **3**, 36 (1965)
19. *Food Standards Committee Report on Vinegar*, H.M.S.O., London (1971)
20. WHITMARSH, J. M., *Analyst, Lond.*, **67**, 188 (1942)
21. ILLING, E. T. and WHITTLE, E. G., *Analyst, Lond.*, **64**, 329 (1939)
22. LYNE, F. A. and MCLACHLAN, T., *Analyst, Lond.*, **71**, 203 (1946)
23. WHITE, J., *Proc. Biochem.*, **6**, No. 5, 21 (1971)
24. ASSOCIATION OF OFFICIAL ANALYTICAL CHEMISTS, *Official Methods of Analysis of the AOAC*, 11th edn, AOAC, Washington, U.S. (1970)
25. WILLIAMS, H. A., *J. Assoc. Publ. Analysts*, **6**, 69 (1968)
26. ASSOCIATION OF PUBLIC ANALYSTS, *J. Assoc. Publ. Analysts*, **9**, 104 (1971)
27. SIPPLE, H. L., *Fd Res.*, **1**, 145 (1936)
28. BLANCHFIELD, J. R., *Food Industries Manual*, 20th edn, Leonard Hill, London, 422 (1969)
29. DICKINSON, D. and GOOSE, P., *Laboratory Inspection of Canned and Bottled Foods*, Blackie, London (1955)
30. INTERNATIONAL FEDERATION OF FRUIT JUICE PRODUCERS, *Analyses*, No. 12, 1 (1964)
31. ROMANN, E. and STAUB, M., *Mitt. Geb. LebensmittelUnters. Hyg.*, **52**, 44 (1961)
32. LIEBMANN, H. and AYRES, A. D., *Analyst, Lond.*, **70**, 411 (1945)
33. KUM-TUTT, L. and LEONG, P. C., *Analyst, Lond.*, **89**, 674 (1964)

10

MISCELLANEOUS

Hardness of Water—Free Chlorine—Alcohol-Insoluble Solids—Caffeine—Assessment of Cocoa Content—Assessment of Egg Content—Solubility of Dried Egg—Creatine and Creatinine in Meat Extract—Jelly Strength of Gelatine —Pesticides

HARDNESS OF WATER

1. EDTA METHOD

METHOD 10.1. ESTIMATION OF TOTAL HARDNESS OF WATER USING EDTA

Reagents

EDTA solution Dissolve 4·0 g of disodium ethylenediaminetetraacetate dihydrate in about 800 ml of water. Add 21·5 ml of 1 N NaOH and 0·1 of $MgCl_2 \cdot 6H_2O$. Standardise the solution by the following method.

Pipette 10 ml of standard calcium chloride solution into a porcelain basin, add 40 ml of water, 0·5 ml of buffer solution and 5 drops of indicator and titrate as described under 'Procedure'. Adjust the EDTA solution so that 1 ml $\equiv$ 1 mg of $CaCO_3$. The end-point is less precise when Mg^{2+} ions are absent. Such precision can be improved by firstly titrating a small volume of tap water to the end-point before adding the 10 ml of calcium chloride solution.

Standard calcium chloride solution Weigh out 1·000 g of pure $CaCO_3$ in a 250-ml beaker and dissolve it in the minimum volume of dilute HCl (added carefully to avoid spurting). Heat the solution gently to boiling, exactly neutralise it with dilute NaOH, transfer it down a rod into a 1-litre volumetric flask and dilute to the mark.

1 ml of solution $\equiv$ 1 mg of $CaCO_3$

Buffer solution Dissolve 40·0 g of borax in 800 ml of water. Dissolve 10·0 g of NaOH and 5·0 g of sodium sulphide in 100 ml of water, cool the solution, mix it with the borax solution and make up the volume to 1 litre with water.

Indicator solution Dissolve 0·50 g of Eriochrome black T in a 3 + 1 v/v mixture of triethanolamine and alcohol.

Procedure

Pipette 50 ml of water sample (with a water of hardness above 200 ppm, take 25 ml of sample plus 25 ml of distilled water) into a porcelain basin, add 0·5 ml of buffer solution by pipette, mix with a rod and add 5 drops of indicator solution. Titrate with the EDTA solution with constant stirring until the colour changes from red to blue (no red or purple colour should be present at the end-point). It is important that the last few drops of EDTA are added slowly with thorough stirring.

$$\text{Total hardness (as } CaCO_3\text{) in ppm} = \frac{\text{Titre} \times 1000}{\text{ml of sample taken}}$$

The hardness is also expressed as mg/l, which is numerically equal to ppm. Waters can be roughly classified as shown in *Table 10.1*.

Table 10.1. CLASSIFICATION OF HARDNESS OF WATERS

Description of hardness	Hardness, mg/l = ppm
Soft	Up to 50
Moderately soft	50–100
Slightly hard	100–150
Moderately hard	150–200
Hard	200–300
Very hard	Over 300

About 40% of the population in the U.K. are supplied with water in the hard caregory of 200–300 mg/l. The hardness of the water used in manufacture affects the quality of many food products.

II. PFEIFER-WARTHA METHOD

METHOD 10.2. ESTIMATION OF TEMPORARY HARDNESS OF WATER BY TITRATION WITH ACID

Titrate 200 ml of water sample with 0·1 N sulphuric acid using methyl orange or benzene-azo-α-naphthylamine as indicator.

$$1 \text{ ml of } 0·1 \text{ N } H_2SO_4 \equiv 0·005 \text{ g of } CaCO_3$$

$$\text{Temporary hardness (as } CaCO_3\text{) in ppm or mg/l} = \frac{\text{Titre}}{0·04}$$

The result can also be referred to as the 'carbonate hardness' or alkalinity, which represents the proportion of the total hardness that is destroyed by boiling.

METHOD 10.3. ESTIMATION OF PERMANENT HARDNESS OF WATER BY TITRATION WITH ACID USING THE PFEIFER–WARTHA METHOD

Mark two 500-ml conical flasks on the outside at 130 ml with a grease pencil. Measure into one flask (A) 200 ml of the water sample. Measure out 200 ml of distilled water into the second flask (B) and then treat both flasks in the following way.

Boil the water for 15 min to expel most of the carbon dioxide. Then add 25 ml of 0.1 N NaOH and 25 ml of 0.1 N sodium carbonate solution (both solutions must be added by separate pipettes). Mix by swirling and boil the solution down until the volume is about 130 ml. Cool the solution, then transfer it with the aid of a funnel into a 200-ml volumetric flask. Wash it in with several *small* volumes of distilled water and make up the volume to the mark. Allow the solution to stand overnight. In the morning, mix thoroughly and filter the solution, rejecting the first 20 ml of filtrate. Pipette 100 ml of filtrate into an evaporating basin or beaker and titrate it with 0.1 N sulphuric acid using methyl orange as indicator.

If the sample titration (A) is a ml and the blank titration (B) is b ml, then permanent hardness (as $CaCO_3$) = 50 $(b-a)$ ppm or mg/l.

With an alkaline water there is no permanent hardness. This would be indicated if $(b-a)$ is negative. If this is so, $(a-b)$ represents the excess of alkali in the water and is usually calculated as sodium carbonate (Na_2CO_3). Also, the alkalinity is in the temporary hardness titration and the excess alkalinity titration should be deducted from the temporary hardness titration to obtain the true temporary hardness.

The permanent hardness is due mainly to the sulphates of calcium and magnesium and is unaffected by boiling. Therefore, from the two chemical titration methods above, the sum of the temporary and permanent hardness gives the total hardness.

FREE CHLORINE

Water is chlorinated in the food industry, the commonest example being the use of about 1 ppm of free chlorine for cooling waters and continuous pasteurisers for cans. Most methods for testing for residual chlorine were devised originally for mains water. Formerly, free chlorine was almost invariably determined colorimetrically by using *o*-tolidine, but the use of this reagent is now discouraged because it is believed to have carcinogenic properties. The following is a simple field test described by Palin[1], in which *p*-aminodimethylaniline is used. As the pink colour produced has a similar tint to that given by nitrites with the Griess–Ilosvay reagents, it can be readily matched against the intensities on the Nessleriser nitrite disc.

METHOD 10.4. ESTIMATION OF FREE CHLORINE IN WATER USING *p*-AMINODIMETHYLANILINE[1]

Apparatus

BDH Lovibond Nessleriser fitted with disc normally used for determining nitrites.

Reagents

Indicator solution Prepare a 0·2% wv solution of *p*-aminodimethylaniline in alcohol. The reagent should preferably be freshly prepared, but can be stored under refrigeration for a few days in an amber bottle.
0·5 M *phosphate buffer solution, pH* 6·8 Dissolve 35·5 g of anhydrous sodium phosphate (Na_2HPO_4) and 34·0 g of potassium dihydrogen phosphate (KH_2PO_4) in water and dilute the solution to 1 litre.

Procedure

Place 50 ml of the water sample in the left-hand compartment of the Nessleriser fitted with the nitrite disc. Into the other glass, pour 50 ml of the water sample, add 1 ml of buffer solution and 0·25 ml of indicator, mix and place the mixture in the right-hand compartment. Match the pink colour produced against the intensities on the nitrite disc. Divide the number appearing on the disc by four to give the free chlorine in the water in ppm (assuming that 50 ml of sample is used). For the chloramine-chlorine, add a crystal of KI and assess the colour increase again from the disc.

Special Nessleriser and comparator discs are also available for use with a further method due to Palin in which diethyl-*p*-phenylenediamine (DPD) is used[2]. Various methods for determining residual chlorine have been evaluated by Nicholson[3].

ALCOHOL-INSOLUBLE SOLIDS

METHOD 10.5. ESTIMATION OF ALCOHOL-INSOLUBLE SOLIDS (AIS) IN CANNED VEGETABLES

Set up a reflux apparatus with a 250-ml flask and a water condenser. Also, dry a 9-cm Whatman No. 1 filter-paper (A) in an oven at 100 °C, preparatory to cooling it in a desiccator and weighing it in a stoppered weighing bottle (cf., Winter[4]).

Macerate the drained sample, weigh 10·0 g of it into a beaker and transfer it to the 250-ml flask with 150 ml of alcohol (80%) with the assistance of a rod. Boil the mixture gently under reflux for 30 min, then filter it through filter-paper A using a Buchner funnel, washing it through with 80% alcohol until the washings are clear and colourless. Dry the paper at 100 °C for 2 h, cool it in a desiccator and weigh it again in the stoppered weighing bottle. Express the AIS as a percentage of the drained sample.

The above method is more usually used for canned peas. In general, smooth-skin varieties give a higher AIS than the wrinkled varieties. Also, canned processed peas (prepared by drying and soaking before canning) give much higher AIS values than canned garden peas (prepared by canning the fresh pea)[5]. The difference appears to be related to the increase found in the AIS as peas become more mature (relatively immature peas are used for the canning of garden peas).

CAFFEINE

The only common food materials that contain caffeine in appreciable amounts are tea, coffee and kola nut. The amount of such materials used in the preparation of products can therefore be assessed approximately from the caffeine content.

METHOD 10.6. ESTIMATION OF CAFFEINE IN LIQUID COFFEE EXTRACTS AND OTHER PRODUCTS

Transfer 20 ml (modified amounts for other coffee products are given at the end of this method) of liquid coffee essence or liquid extract to a 100-ml volumetric flask and add 60 ml of water. Add 5 ml of zinc acetate solution, mix, add 5 ml of potassium ferrocyanide and mix again (see p. 60). Make up the volume to the mark with water, mix, allow the mixture to stand for at least 5 min, then filter it through a rapid filter-paper into a 100-ml graduated cylinder. Measure the volume (V ml) of filtrate and pour all of this into a 200-ml separator. Then add 10 ml of ammonia of sp. gr. 0·880 and extract the solution three times with 40-, 30- and 10-ml portions of chloroform, pouring each extract into a further separator. Wash the bulked extracts with 5 ml of 1% aqueous KOH and then with 5 ml of water. Transfer the chloroform extract to a flask and remove the chloroform (*Figure 6.3*) until the volume is 10–15 ml. Transfer the solution with chloroform into a weighed beaker, evaporate it carefully, dry it at 100 °C for 30 min, then cool and weigh it (W g).

$$\text{Caffeine } (\% \text{ w/v}) = \frac{W}{20} \times \frac{100}{V} \times 100$$

The extracted caffeine residue should preferably consist of white crystals. If the residue is coloured, the result is liable to be too high. For more accurate work, re-dissolve the extract in a small volume of chloroform and carefully transfer the solution down a rod into a 100-ml long-necked Kjeldahl flask. Wash it in with small volumes of chloroform and then remove the chloroform by immersing the flask in boiling water. Add 2 g of potassium sulphate, 0·2 g of copper sulphate, 20 ml of concentrated sulphuric acid and a few glass balls, and digest the mixture for at least 45 min. Cool and dilute the mixture, transfer it to a distillation apparatus and continue as for the macro Kjeldahl method (*Method 2.6a*, p. 52).

1 ml of 0·1 N H_2SO_4 ≡ 4·855 mg of anhydrous caffeine
% N × 3·464 = anhydrous caffeine

Alternatively, the caffeine in chloroform solution can be determined spectrophotometrically ($E_{1\text{ cm}}^{1\%}$ = 532 at 273 nm).

The statutory minimum for caffeine in liquid coffee extract is 0·50% w/v (0·25 for liquid coffee and chicory extract and 0·40% for liquid extract of coffee and fig).

Solid coffee extract (*instant coffee*)

Dissolve 5 g of instant coffee in hot water, transfer the solution to a 100-ml volumetric flask, add the zinc ferrocyanide clearing agents and continue as for liquid extracts. Instant coffees contain 2·5–8% of caffeine (there is no statutory minimum).

Tea, coffee, coffee and chicory mixture

Boil 2 g sample under reflux with at least 250 ml of water. Filter the solution, wash it with hot water and evaporate it to about 40 ml. Transfer it to a 100-ml volumetric flask, add the zinc ferrocyanide clearing agents and continue as for liquid extracts. Tea and coffee contain about 2–4% and 1·0–1·3% of caffeine, respectively.

ASSESSMENT OF COCOA CONTENT

Cocoa contains a relatively high proportion of 'total alkaloids'. This content, consisting of theobromine and caffeine, is therefore used for assessing the amount of cocoa present in products such as chocolate cake, cake mix and chocolate itself. The following spectrophotometric method due to Chapman et al.[6] is more rapid than that described by Moir and Hinks[7], which was invariably used previously.

METHOD 10.7. SPECTROPHOTOMETRIC ESTIMATION OF TOTAL ALKALOIDS IN COCOA AND ITS PRODUCTS

Principle of method

The 'alkaloids' are adsorbed on Fuller's earth and then removed with alkali. After acidifying, the UV absorption is measured at 270–275 nm.

Reagents

The only special reagent necessary is the Fuller's earth, which should be of a special grade suitable for the adsorption. In view of its variable nature, each batch should be tested before use. '

Procedure

Weigh accurately 1 g of cake-mix or powdered or minced flour confectionery or 0·1 g of cocoa into a 100-ml beaker. Mix to a smooth paste with 25 ml of water, added in small portions. Heat the mixture gently, stirring continuously, until the liquid boils and then place it on a boiling water bath for 15 min. Wash the mixture into a 50-ml volumetric flask with a small volume of water, cool it and add by separate pipettes 1 ml of each of the zinc ferrocyanide clearing agents (p. 60), gently mixing after each addition. Dilute the solution to the 50-ml mark and mix thoroughly. Then centrifuge the solution in a suitable tube at 2500 rpm for 5 min and decant it through a Whatman No. 4 filter-paper into a dry beaker. Weigh 0·5 g of selected Fuller's earth into a dry centrifuge tube, add by pipette 30 ml of filtrate and mix well with a rod. As the rod is removed from the tube, wash it with a few millilitres of water. Centrifuge the mixture at 2500 rpm for at least 5 min and pour off the clear supernatant liquid without disturbing the residue. Then add 15 ml of water to the residue, and as before, mix, rinse, centrifuge and discard the clear liquid.

Transfer the residue to a 50-ml volumetric flask with 0·1 N NaOH, dilute the solution to the mark with 0·1 N NaOH and shake it hard for 2 min. Then centrifuge it at 2500 rpm for at least 10 min and decant the clear liquid carefully into a beaker. Pipette a 30-ml aliquot of this liquid into a 100-ml volumetric flask, add 13 ml of 1 N HCl, dilute the solution to the mark with water and mix. Measure the UV absorption at 270, 271, 272, 273, 274 and 275 nm against distilled water and also that of a blank containing 30 ml of 0·1 N NaOH + 13 ml of 1 N HCl made up to 100 ml. Theobromine has a maximum absorption $(E_{1\,cm}^{1\%})$ of 548 at 272 nm and the sample should show a peak at 272 ± 1 nm for the assay to be valid. If E is that obtained for a 1-cm cell at 272 nm, then

$$\text{Total alkaloids (\% w/w)} = \frac{(E_{sample} - E_{blank}) \times 50 \times 50 \times 100}{548 \times 30 \times 30 \times \text{Wt. of sample taken}}$$

$$= \frac{(E_{sample} - E_{blank}) \times 0\cdot51}{\text{Wt. of sample taken}}$$

$$\text{Average dry fat-free cocoa (\%)} = \text{Total alkaloids (\%)} \times 33\cdot3$$

According to LAJAC Code of Practice No. 1, where the word 'chocolate' is used in the description of flour confectionery, the product must contain at least 3% of dry non-fat cocoa in the moist crumb.

ASSESSMENT OF EGG CONTENT

Egg contains a relatively high proportion of organic phosphorus. The proportion of egg in products can therefore be assessed from the percentage of phosphorus (as P_2O_5) that can be extracted by an organic solvent. Manley and Lobley[8] have shown that the amount of phosphorus that can be extracted from dried egg varies according to the solvent used. In the following method 95% alcohol is used, but the technique is the same for all solvents.

METHOD 10.8. ESTIMATION OF EGG CONTENT OF PRODUCTS SUCH AS SALAD CREAM AND FRUIT CURD[9]

Weigh 20 g of sample such as salad cream, lemon curd or cake into a flask, add 100 ml of 95% alcohol and attach a water reflux condenser. Heat the mixture in a boiling water bath for 6 h and allow it to stand overnight. Filter it through a Buchner or, preferably, a Hartley funnel (*Figure 10.1*). Wash the residue with a small volume

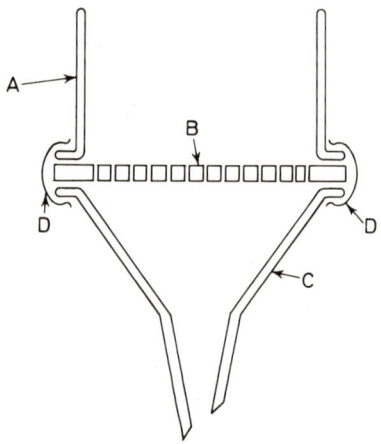

Figure 10.1. Hartley modification of Buchner funnel (three-section pattern).
The filter-paper is placed across the perforated plate, B. After placing A above and C below the plate, the funnel is sealed tight by clips, D

of 95% alcohol. Retain the filtrate (A). Return the residue in the funnel to the flask and re-extract it by heating it under reflux again with 100 ml of 95% alcohol. Then filter the mixture as before and add the filtrate to the previous filtrate A.

Place a silica basin on a boiling water bath and pour the combined filtrate into it in small portions until it is evaporated to dryness. Gently ignite the residue, then warm and stir the ash with 5 ml of dilute nitric acid, add 5 ml of water and filter the mixture into a 100-ml volumetric flask, B. Wash the filter with a small volume of water. Return the filter-paper to the dish, ignite it, then warm and stir the ash with 2 ml of dilute nitric acid, add 3 ml of water, filter the mixture into B and wash the filter with a small volume of water. Dilute the cooled combined filtrate with water to the 100-ml mark, mix and pipette 50 ml of the solution into another 100-ml volumetric flask. Add a small piece of indicator paper, neutralise the solution with ammonia of sp. gr. 0·880, make it just acidic with dilute nitric acid, add 25 ml of vanadomolybdate reagent, dilute the solution to the mark with water, mix and measure the optical density at 420 nm as described in *Method 4.7b* (p. 110). If, as is usual, the reading falls outside the range of the calibration graph, take a different volume of solution for the determination. Alternatively, other procedures for determining the phosphate, such as the phosphomolybdate volumetric method, *Method 4.7a* (p. 108). can be used. First, calculate the percentage of phosphorus (as P_2O_5) extracted from the original sample and assess the amount of egg present in the product from the following equations:

$$\text{Dried egg } (\%) \quad = \% \, P_2O_5 \times 83$$
$$\text{Egg-yolk solids } (\%) = \% \, P_2O_5 \times 58$$

Allowances must be made for other ingredients that contain extractable phosphorus. Thus, whilst 95% alcohol extracts 1·15–1·25% of P_2O_5 from dried egg, it also extracts 0·08% from wheat flour and 0·01% from cornflour.

Statutory regulations require salad cream to contain at least 1·35% w/w of egg-yolk solids and lemon curd at least 1% w/w of dried whole egg. In 1968, the FSC recommended that the minimum standard for egg in fruit curds should be expressed as 0·7% egg-yolk solids.

Other techniques for estimating egg have been described by the AOAC (the digitonin method—see also Stauffer[10]) and by Casson and Griffin[11, 12] (from the amount of choline liberated).

SOLUBILITY OF DRIED EGG

The solubility of dried egg is related to its physical deterioration during drying or storage. Hawthorne[13] has reviewed various methods for its determination, in which the sample is treated with water or aqueous reagents and either (*a*) the dissolved solids are weighed or (*b*) the solubility of a particular constituent is measured. In the Haenni method, the solubility is assessed from the change in the refractive index of a 5% salt solution after shaking it with a sample of dried egg.

METHOD 10.9. ESTIMATION OF THE SOLUBILITY OF DRIED EGG BY THE MODIFIED HAENNI METHOD[13]

Weigh 1·0 g of dried egg into a test-tube and add 5 ml of 5% w/v sodium chloride solution. Close the tube with a rubber stopper, disperse the powder by shaking the tube gently for 1 min, allow it to stand for 15 min, then invert it 10 times. Thirty minutes after the initial shaking, invert the tube again 10 times. After a further 5 min, close a piece of glass tubing of 2-mm bore with the index finger, insert this 'pipette' just below the surface of the liquid in the test-tube and rotate it sharply. Release the finger momentarily, remove the 'pipette' from the solution, wipe the outside of it and transfer a drop of the liquid on to the prism of a refractometer

$$\text{Solubility} = \frac{\log_{10}\left[1000(N_D^{25} \text{ of sample} - N_D^{25} \text{ of solvent})\right] - 0\cdot445}{0\cdot01}$$

where N_D^{25} denotes the refractive index of the salt extract.
Commercial samples of dried egg have solubilities of about 50–99%. The solubility gradually falls during storage. When a sample of high solubility deteriorates, in the initial stages a small decrease in soluble nitrogen is accompanied by a large drop in refractive index.

CREATINE AND CREATININE IN MEAT EXTRACT

METHOD 10.10. DETERMINATION OF TOTAL CREATINE AND CREATININE IN MEAT EXTRACT[14]

Procedure

Weigh out 10·0 g of sample into a 100-ml beaker, add water, stir the mixture thoroughly with a rod and transfer it carefully down the rod into a 100-ml volumetric flask. Dilute the solution to the mark and mix. Either heat 10 ml of the 10% solution with 10 ml of 2 N HCl in boiling water under a water reflux condenser for 2–3 h, or auto-clave this mixture for 20 min at 120 °C. Cool the hydrolysate, add

10 ml of 2 N NaOH and dilute the solution to 500 ml. Mix and pipette 5 ml of the solution into a 100-ml volumetric flask and add about 15 ml of water, 20 ml of 1 % aqueous picric acid and 2·5 ml of 2 N NaOH. After 15 min, dilute the mixture to 100 ml with water, mix and filter it, rejecting the first few millilitres. Measure the orange colour in a 1-cm cell of a spectrophotometer at 520 nm. Similarly, measure the colour of a reagent blank. By comparing the readings against the reference graph, obtain the amount of creatine plus creatinine in the sample as creatinine.

Preparation of reference graph

Dissolve 1·603 g of creatinine zinc chloride in 0·1 N HCl and make up the volume to the mark in a 1-litre volumetric flask with 0·1 N HCl. Then dilute the solution ten times so that 1 ml of dilute solution (A) ≡ 0·1 mg of creatinine. Into a series of 100-ml volumetric flasks, pipette 0, 2, 4, 6, 8 and 10 ml of standard solution A (≡ 0–1·0 mg of creatinine). Add water to each flask to bring the volume to 20 ml, then add 20 ml of 1 % picric acid and 2·5 ml of 2 N NaOH. After 15 min, dilute each solution to 100 ml, mix and measure the colour at 520 nm as for the sample. Prepare the reference graph covering the range 0–1·0 mg of creatinine (corresponding to the amounts present in 100 ml).

The amount of creatinine present in concentrated meat extract should not fall below about 10 % when calculated on the dry matter. Products of the Bovril type contain about 1·3 % of creatine plus creatinine.

JELLY STRENGTH OF GELATINE

In the food industry, the estimation of the jelly strength of gelatine in absolute terms is seldom required. More usually, the results are compared with those obtained on a reference sample retained from a previous delivery (preferably the first). As the jelly strength of gelatine varies according to various factors, it is preferable to set the solution under conditions approximately the same as those under which it will be set in the final made-up product, e.g., at the same concentration and pH, and after heating the solution in boiling water. The jelly strength of gelatine gels can be assessed by the finger or by using commercial instruments. An example of each of these methods is given below.

281

METHOD 10.11. ESTIMATION OF THE JELLY STRENGTH OF GELATINE

Reference sample

Store in a stoppered jar a reasonable amount of suitable gelatine for the particular manufacturing process, preferably taken from the first delivery (R) (pp. 5–7).

N.B. In the following methods, it is assumed that the gelatine would be set in the final product (or final made-up jelly in the home) at a concentration of about 5%. The amounts can be suitably modified for other circumstances, including pH adjustment.

Procedure A (finger method)

Obtain a series of similar straight-sided tumblers or other suitable containers, preferably with a total capacity of about 125 ml.

Weigh out 5 g of powdered or kibbled sample (G) into one container, add 100 ml of tap water, allow the sample to soak for at least 1 h (3 h if kibbled) and then warm it in a water-bath at 70 °C with constant gentle stirring to effect dissolution. As heat causes a loss in jelly strength, the temperature of the solution should always remain below 60 °C.

At the same time, weigh out the following amounts of reference sample R into similar vessels: 4·5, 4·75, 5·0, 5·25 and 5·5 g. Soak each sample and heat it with 100 ml of tap water to effect dissolution, as for G. These weights correspond to 90, 95, 100, 105 and 110% of the 'standard', respectively.

Cool all the solutions, which should contain no undissolved particles, in cold water and incubate them overnight at the agreed temperature (preferably 10 °C if a suitable incubator or controlled room is available). In the morning, compare the jelly strengths of the set jellies by pushing the index finger into them. By 'interpolation', estimate the jelly strength of G as a percentage of the standard, S.

Procedure B (Boucher Jelly Tester)

The construction of the Boucher Jelly Tester is described by Koprowski[15].

Place 5·0 g of powdered or kibbled sample in the test bottle, add 100 ml of tap water and allow the sample to soak for at least 1 h (3 h if kibbled). Then warm it in a water-bath at 70 °C with constant gentle stirring to effect dissolution, so that the solution temperature does not exceed 60 °C. Check that there are no undissolved particles present, cool the bottle in water, insert the rubber stopper and incubate the mixture overnight at 10 °C. In the morning remove the stopper and transfer the sample bottle to the table of the instrument immediately after its removal from the 10 °C incubator. Raise the table until (a) the plunger just comes into contact with the set jelly surface and (b) the zero-indicating line is in the centre of the microscope-type zero finder. Then open the needle valve at the top so that the tap water in the upper reservoir runs into the bucket suspended from the balance beam. The weight of this water is transmitted through the beam to the plunger, which presses on to the jelly. When the plunger is depressed 5 mm into the jelly surface, the flow of water into the bucket stops automatically. When this occurs, remove the bucket and measure the volume of water it contains, in millilitres, by pouring the water into a suitable measuring cylinder.

Preparation of reference graph

Into a series of bottles weigh out the following amounts of reference sample R: 4·5, 4·75, 5·0, 5·25 and 5·5 g, and treat them as for the sample. These weights correspond to 90–110% of the 'standard'. Construct the reference graph relating jelly strength (as millilitres of water) and the percentage of 'standard' gelatine. Repeat the determination on 5·0 g of R with every subsequent sample examined. If necessary, any variation in the figure obtained should be allowed for in assessing the jelly strength of the sample.

Interpretation of results

By comparison with the reference graph, express the jelly strength as a percentage of the standard gelatine and relate it if necessary to the recipe weight per batch.

Alternatively, carry out the method on a 6·66% (instead of 5%) solution and call the water volume (millilitres) 'Boucher jellograms'. Then:

$$\text{Bloom degrees} = \left(\text{Boucher jellograms} \times \frac{2}{3}\right) + 8$$

Gelatine is sold commercially on the basis of the Bloom value, which can be obtained directly from the Bloom Gelometer, the operation of which is described in British Standard 757:1959.

Gelatine with a jelly strength of 230° Bloom is normally suitable for the manufacture of table jellies. The statutory requirement for the jelly strength of gelatine is that a 3% solution shall set to a jelly when it is cooled to and maintained at 15·5 °C.

PESTICIDES

Residues of various pesticides may be found in foods, particularly after spraying crops of fruits and vegetables. In view of their toxic effects, it is advisable to monitor the amounts present in raw materials. Where possible the amount and type of treatment to be applied should also be laid down in specifications agreed between farm and factory. In view of their persistency and toxicity the most important group are the organochlorine insecticides such as DDT, BHC, Aldrin, Dieldrin and Heptachlor. Although the organo-phosphorus insecticides such as Malathion are the more toxic they tend to be rapidly metabolised to less toxic residues. Other materials of importance are the dithiocarbamates. In 1967 the FACC recommended that limits for aldrin and dieldrin should be prescribed for certain foods.

Details of the methods used for examining foods for pesticides are outside the scope of this volume. In outline, these involve extraction of the pesticide from the sample, removal of interfering substances and the identification and estimation of the amount present. Schemes for the extraction and clean-up have been described for vegetables[16, 17] and animal[18] products. Hexane and acetone are frequently used for the extraction and sodium sulphate for the removal of water.

Pesticides are usually identified and estimated by means of TLC or GLC (*Table 10.2*). For the GLC techniques, which are considered to be the most accurate and efficient for the purpose, Celite 545 and Chromosorb G are commonly employed as solid supports, and silicone elastomers, Epikote resin and Apiezon grease as the stationary phases. Bio-assay methods using flies are also useful as a sorting test (Hall[19]).

Table 10.2. REFERENCES TO METHODS FOR PESTICIDES

Pesticide class	TLC methods	GLC methods
Organochlorine	Abbott et al.[20]	Dickes & Nicholas[21] Hamence et al.[22]
Organophosphorus	Bunyan[23]	Dickes & Nicholas[24] Egan et al.[25]
Dithiocarbamate fungicides	Pagington[26]	—

MISCELLANEOUS

REFERENCES

1. PALIN, A. T., *Analyst, Lond.*, **70**, 203 (1945)
2. PALIN, A. T., *Proc. Soc. Water Treat. Exam.*, **6**, No. 2, 133 (1957)
3. NICOLSON, N. J., *Analyst, Lond.*, **90**, 187 (1965)
4. WINTER, F. H., *J. Ass. off. analyt. Chem.*, **54**, 54 (1971)
5. HOULBROOKE, A., *J. Assoc. Publ. Analysts*, **1**, 16 (1963)
6. CHAPMAN, W. B., OGDEN, E. and URRY, S., *J. Assoc. Publ. Analysts*, **1**, 59 (1963)
7. MOIR, D. D. and HINKS, E., *Analyst, Lond.*, **60**, 439 (1935)
8. MANLEY, C. H. and LOBLEY, H., *Analyst, Lond.*, **73**, 30 (1948)
9. BAGNALL, D. J. T. and SMITH, A., *Analyst, Lond.*, **70**, 211 (1945)
10. STAUFFER, L. J., *J. Ass. off. analyt. Chem.*, **50**, 851 (1967)
11. CASSON, C. B. and GRIFFIN, F. J., *Analyst, Lond.*, **84**, 281 (1959)
12. CASSON, C. B. and GRIFFIN, F. J., *Analyst, Lond.*, **86**, 544 (1961)
13. HAWTHORNE, J. R., *J. Soc. chem. Ind., Lond.*, **63**, 6 (1944)
14. SOCIETY OF PUBLIC ANALYSTS, *Analyst, Lond.*, **76**, 329 (1951)
15. KOPROWSKI, W. S., *Analyst, Lond.*, **76**, 732 (1951)
16. LAWS, E. Q. and WEBLEY, D. J., *Analyst, Lond.*, **86**, 249 (1961)
17. GOODWIN, E. S., GOULDEN, R. and REYNOLDS, J. G., *Analyst, Lond.*, **86**, 697 (1961)
18. JONES, L. R. and RIDDICK, J. A., *Analyt. Chem.*, **24**, 569 (1952)
19. HALL, P. S., *J. Assoc. Publ. Analysts*, **1**, 5 (1963)
20. ABBOTT, D. C., EGAN, H. and THOMSON, J., *J. Chromatog.*, **16**, 481 (1964)
21. DICKES, G. J. and NICHOLAS, P. V., *J. Assoc. Publ. Analysts*, **5**, 52 (1967)
22. HAMENCE, J. H., HALL, P. S. and CAVERLY, D. J., *Analyst, Lond.*, **90**, 649 (1965)
23. BUNYAN, P. J., *Analyst, Lond.*, **89**, 615 (1964)
24. DICKES, G. J. and NICHOLAS, P. V., *J. Assoc. Publ. Analysts*, **6**, 60 (1968)
25. EGAN, H., HAMMOND, E. W. and THOMSON, J., *Analyst, Lond.*, **89**, 175 (1964)
26. PAGINGTON, J. S., *J. Assoc. Publ. Analysts*, **6**, 25 (1968)

APPENDIX I

NOTES ON FOOD LEGISLATION

FOOD AND DRUGS ACT 1955

The consumer is protected as to the composition and labelling of food by the basic provisions in the Food and Drugs Act. The operation of the Act is the responsibility of the Food and Drugs Authorities, which consist essentially of the County, Borough and Rural District Councils, who appoint a sampling officer to procure the samples and a public analyst to test them.

Most prosecutions are made under Section 2, which prohibits the sale of food that is not of the nature, substance or quality demanded, if this is to the prejudice of the purchaser. Also of considerable importance is Section 8, which prohibits the sale of food that is unfit for human consumption. Summonses under the Act can be taken out against any person or body considered to be responsible, including the manufacturer, packer, labeller or retailer.

The Minister of Agriculture, Fisheries and Food is given power to make regulations as to the composition and labelling of food and the control of additives and contaminants. These are made under Statutory Instruments (S.I.s), which can be purchased from H.M.S.O. (*Table A.1*). The Secretary of State for Social Services also has certain responsibilities under the Act.

Table A.1. LIST OF MAIN REGULATIONS MADE UNDER OR PERPETUATED BY THE FOOD AND DRUGS ACT 1955

Subject of regulation	SR & O/SI, Year and No.	Type of main provisions at January, 1972[a]
Antioxidants	1966, 1500	Gallates, BHA and BHT permitted in oils and fats
Arsenic	1959, 831; 1960, 2261	Arsenic limits for foods. General max. 1·0 ppm As

286

Subject of regulation	SR & O/SI, Year and No.	Type of main provisions at January, 1972[a]
Baking powder	1944, 46; 1946, 157	Min. available CO_2 8%, max. residual CO_2 1·5%
Bread	1963, 1435	Definitions and standards for breads
Butter	1966, 1074	Min. 80% milk fat, max. 16% water, max. 2% MSNF
Cheese	1970, 94	Min. fat, max. water according to variety
Coffee and chicory	1967, 1865	Min. 51% coffee
Coffee essences	1967, 1865	Min. caffeine from 0·25 to 0·50%
Coffee, instant	1967, 1865	Min. 95% soluble solids from coffee
Colours	1966, 1203; 1970, 1102	Permitted list of added colours
Condensed milk	1959, 1098	Min. fat, min. TMS according to product
Cream	1970, 752	Min. milk fat from 12 to 55%
Curry powder	1949, 1816; 1956, 1166	Min. 85% spices, max. 20 ppm Pb
Dried milk	1965, 363	Min. fat according to product, max. 5% moisture
Egg, liquid	1963, 1503	Compulsory heat treatment and test for adequacy
Emulsifiers and stabilisers	1962, 720; 1970, 1101	Permitted list of emulsifiers and stabilisers
Fish cakes	1950, 589	Min. 35% fish
Fish paste	1968, 430	Min. fish for fish paste (70%), potted fish, etc.
Flour	1963, 1435	Standards for nutrients
Flour, self-raising	1946, 157	Min. available CO_2 0·40%
Fluorine	1959, 2106	Max. fluorine in baking powder, etc.
Fruit curd	1953, 691	Min. fat, citric acid, volatile oils, egg, soluble solids
Gelatine	1951, 1196 and 2240	Max. ash, As, Cu, Pb, Zn
Golden raising powder	1944, 46; 1946, 157	Min. available CO_2 6%, max. residual CO_2 1·5%
Hygiene, food	1970, 1172	Sanitary conditions of premises, equipment, etc.
Ice cream	1967, 1866; 1959, 734; 1963, 1083	Compositional minima, e.g., for fat and MSNF
Irradiation	1967, 385; 1969, 1039	Max. for residual ionising radiation
Jam and marmalade	1953, 691 and 1307	Min. for soluble solids and fruit
Labelling	1970, 400	Labels to bear manufacturer's name and address, list of ingredients present, etc.
Lead	1961, 1931	Lead limits for foods. General max. 2 ppm Pb
Margarine	1967, 1867	Min. fat. Max. for butter fat and water. Vitamins A and D to be added

Subject of regulation	SR & O/SI, Year and No.	Type of main provisions at January, 1972[a]
Mayonnaise	1966, 1051	Min. 25% oil, 1·35% egg-yolk solids
Meat, canned	1967, 861; 1968, 2046	Min. for meat from 15% to 'wholly meat'
Meat paste	1968, 430	Min. meat for meat paste (70%), potted meat, chopped meat, etc.
Meat pies	1967, 860	Min. meat content, e.g., cooked meat pie 25%
Meat treatment	1964, 19	Addition of vitamin C and nicotinic acid to raw meat prohibited
Milk	1939, 1417	Min. 3% fat and 8·5% NFS (both presumptive)
Milk, Channel Island and South Devon	1956, 919	Min. 4% fat
Milk, pasteurised, sterilised and UHT	1963, 1571; 1965, 1555	Heat treatment requirements and tests for adequacy
Milk, filled	1960, 2331; 1968, 1474	Compositional and labelling requirements for milks containing non-milk fat
Mincemeat	1953, 691	Min. dried fruit, sugar, suet, soluble solids. Max. acetic acid
Mineral oil	1966, 1073	Limits for residual hydrocarbons in a few foods
Mustard	1944, 275; 1946, 157; 1947, 650; 1948, 1073	Min. 0·35% allyl isothiocyanate, max. flour and spices 20%
Preservatives	1962, 1532; 1971, 882	Limits for tolerated preservatives in named foods
Saccharin tablets	1969, 1817	Standards for saccharin in full-strength and half-strength tablets
Salad cream	1966, 1051	Min. 25% oil, 1·35% egg-yolk solids
Sausages	1967, 862; 1968, 2047	Min. meat and lean meat in sausages and other meat products
Sausage rolls	1967, 860	Min. meat content, e.g., 12·5% for cooked products
Soft drinks	1964, 760; 1969, 1818; 1970, 1597	Min. sugar and fruit, max. saccharin for carbonated and concentrated drinks
Solvents	1967, 1582 and 1939	Permitted list of solvents
Suet	1952, 2203	Min. 83% beef fat in shredded suet
Sweeteners, artificial	1969, 1817	Saccharin is the only permitted artificial sweetener
Tomato ketchup	1949, 1817; 1956, 1167	Min. 6% tomato solids, max. 20 ppm Cu

[a] For full details consult the appropriate H.M.S.O. publications.

In the absence of specific statutory standards, the analyst must take the following into account, as appropriate:

(a) Precedents laid down in the higher courts.

(b) Recommendations contained in Reports of the Food Standards Committee and the Food Additives and Contaminants Committee (published by H.M.S.O.). Such recommendations represent advice to appropriate 'Ministers' and often form the basis of future regulations.

(c) Recognised standards, especially those in Codes of Practice formulated by the Local Authorities' Joint Advisory Committee on Food Standards (LAJAC).

(d) The view of the Association of Public Analysts (APA).

(e) Views of trade bodies and authoritative experts.

(f) Overseas standards.

THE WEIGHTS AND MEASURES ACT 1963

The Weights and Measures Act is the responsibility of the county authorities and makes it an offence to sell goods that are short in weight, measure (volume) or number (as with eggs). Wrapper weights are also limited for many foods.

TRADE DESCRIPTIONS ACT 1968

The Trade Descriptions Act is also the responsibility of the weights and measures inspectors appointed by the county authorities. It prohibits the application of false trade descriptions to any goods.

Details of provisions relating to food legislation can be found in the following reference work: W. J. Bell and J. A. O'Keefe, *Sale of Food and Drugs,* 14th edn, and *Service Volume,* Butterworths, London (1968).

Information on new and pending legislation can be acquired by regularly reading as many publications as possible, e.g., the *British Food Journal,* the *Journal of the Association of Public Analysts* and the various trade journals.

APPENDIX II

NOTES ON SPECTROPHOTOMETRY

Light or radiant energy passing through a transparent medium may be absorbed, transmitted or reflected. In absorption spectrophotometry, the amount of absorption is considered.

Wavelength (λ) is expressed in one of the following forms:

$$\text{nanometre (nm)} = \text{millimicron (m}\mu\text{)} = 10^{-6}\text{mm}$$

$$\text{Ångstrom (Å)} = 10^{-7}\text{mm} = \frac{\text{m}\mu}{10}$$

Ranges of wavelength measurements

Ultra-violet	185–400 nm	Use UV spectrophotometer
Visible	400–760 nm	Use absorptiometer with coloured filters (*Tables A.2 and A.3*) or spectrophotometer
Infra-red	0·76–15 μm	Use infra-red spectrophotometer

Table A.2. FILTERS FOR ABSORPTIOMETRY

Wavelength/nm	Colour of filter	Colour observed
400	Violet	Greenish yellow
425	Indigo–Blue	Yellow
450	Blue	Orange
490	Blue–Green	Red
510	Green	Purple
530	Yellow–Green	Violet
550	Yellow	Indigo–Blue
590	Orange	Blue
640	Red	Bluish-green
730	Deep red	Green

Table A.3. TRANSMISSION REGIONS OF ILFORD SPECTRUM FILTERS

Number	Colour of filter	Peak wavelength/nm	Transmission region/nm
600	Spectrum deep violet	405	380–450
601	Spectrum violet	425	380–470
602	Spectrum blue	470	440–490
603	Spectrum blue–green	490	470–520
604	Spectrum green	520	500–540
605	Spectrum yellow–green	550	530–570
606	Spectrum yellow	580	560–610
607	Spectrum orange	600	575 onwards, with absorption increasing from 600
608	Spectrum red	660	620 into infra-red
609	Spectrum deep red	690	650 into infra-red
621	Bright spectrum violet	445	340–515
622	Bright spectrum blue	470	375–530
623	Bright spectrum blue–green	490	460–545
624	Bright spectrum green	520	490–575
625	Brighter spectrum yellow–green	540	510–590
626	Bright spectrum yellow	575	545–620

Lambert's law states that the proportion of radiant energy absorbed by a substance is independent of the incident radiation.

Beer's law states that the absorption is proportional to the total number of molecules of the solute in the path of the radiation.

If these two laws are combined:

$$\text{Extinction} = E = \varepsilon c l = \log_{10} \frac{I_o}{I}$$

where

ε = molecular extinction coefficient
c = concentration of substance (mole per litre)
l = thickness of solution (cm)
I_o = intensity of incident light
I = intensity of transmitted light

In practice, it is more usual to use the extinction coefficient ($E_{1\,cm}^{1\%}$), which is the value of $\log \frac{I_o}{I}$ for a 1-cm path through a 1 % solution of the substance. The formula then becomes:

$$\text{Extinction coefficient} = E_{1\,cm}^{1\%} = \frac{A \times 1(\text{cm}) \times 1(\%)}{l \times c}$$

where

    A = observed absorbance or optical density

    l = cell length (cm)

    c = concentration (%)

The term optical density refers to any extinction measured, irrespective of the concentration and light path. Hence:

$$\text{Optical density} = A = \log \frac{I_o}{I}$$

and

$$\log T = 2 - A$$

where

$$T = \% \text{ transmission} = \frac{I}{I_o} \times 100$$

CALCULATION OF CONCENTRATION FROM THE EXTINCTION COEFFICIENT

From a value for $E_{1\,cm}^{1\%}$ given in a method, the concentration in the solution measured can be obtained by rearranging the above equation:

$$\text{Concentration} (\%) = \frac{A \times 1 \times 1}{l(\text{cm}) \times E_{1\,cm}^{1\%}}$$

$$= \frac{A}{l \times E_{1\,cm}^{1\%}}$$

SETTING UP A REFERENCE GRAPH IN SPECTROPHOTOMETRY

All readings are taken at the same wavelength, usually the wavelength of maximum absorption (*Table A.3*). The controls of the instrument are adjusted to zero absorption on the logarithmic scale (100% on the transmission scale) with either the pure solvent (e.g., water or chloroform) or the reagents as used for the sample (usually 'taking' 0 ml of the standard solution) in the cell. If the pure solvent is used, the reagents in the '0 ml' standard usually give some colour and a small optical density is obtained such as A from AB (*Figure A.1*). A blank (omitting the sample) may also be required. This would involve the use of other reagents prior to the colorimetric reaction, e.g., the acid used to dissolve the ash in trace metal

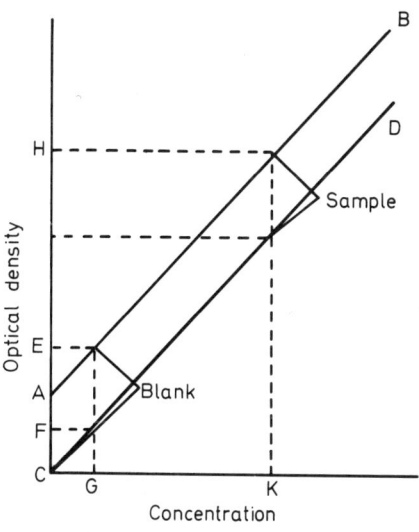

Figure A.1.

determinations. This would give a reading *E*, corresponding to concentration *G*, which is then subtracted from the sample (including blank) concentration *K*, corresponding to reading *H*. If the instrument is set against the '0 ml' standard, the corresponding reference graph is *CD* in which *C* represents zero absorption and reading *F* corresponds to the blank concentration *G*, which has to be subtracted from *K* as before to give the amount present in the sample.

APPENDIX III

FACTORS FOR FOOD ENERGY CALCULATIONS

According to The Labelling of Food Regulations 1970, the following factors should be used:

	Calorie conversion factors, kcal/g
Carbohydrate	3·75
Glycitol	3·75
Protein	4
Alcohol	7
Fat	9

The unit of energy in the SI system is the joule (J)—see p. 302.

Glycitol (or alditol) is a general term denoting sugar alcohols, a group which includes mannitol and sorbitol.

APPENDIX IV

CONTROL OF CARCINOGENIC SUBSTANCES

The Carcinogenic Substances Regulations 1967 apply to factories and premises covered by the Factories Act.

Table A.4. SUMMARY OF THE CARCINOGENIC SUBSTANCES REGULATIONS 1967

Manufacture, presence and uses prohibited:

2-Naphthylamine	Their salts
Benzidine	Substances containing these unless present in very small
4-Aminobiphenyl	concentrations
4-Nitrobiphenyl	

Employment of personnel in manufacture and uses controlled:

1-Naphthylamine	Dichlorobenzidine
o-Tolidine	Their salts
Dianisidine	

Employment of personnel in manufacture controlled:

Auramine
Magenta

APPENDIX V

PREPARATION OF INDICATORS

ACID–BASE INDICATORS

The methods of preparing the main acid–base indicators are outlined in *Table A.5*. In the case of methyl orange, a more sensitive end-point is obtained if it is screened with xylene cyanol as in the following modification:

SCREENED METHYL ORANGE—Dissolve 0·1 g of methyl orange and 0·26 g of xylene cyanol FF in 50 ml of 95 % alcohol and dilute the solution to 100 ml with water.

SCREENED METHYL RED—See p. 52.

SOME INDICATORS FOR OTHER PURPOSES

FERRIC AMMONIUM SULPHATE—Prepare a 10 % w/v aqueous solution.

ACID FERRIC AMMONIUM SULPHATE—Dissolve 0·2 g of ferric ammonium sulphate in 50 ml of water, add 6 ml of dilute nitric acid (concentrated acid diluted 1 + 9 with water) and sufficient water to give a volume of 100 ml.

POTASSIUM CHROMATE—Prepare a 5 % w/v aqueous solution.

STARCH SOLUTION—Mix the minimum volume of water with 1 g of starch (preferably soluble starch) and 5 mg of mercuric iodide to form a smooth paste. Then stir in 500 ml boiling water and boil the mixture for a further 1–2 min. If the solution is not clear, allow it to stand to separate. Decant off the clear supernatant liquid into the indicator bottle.

Table A.5. A RANGE OF ACID–BASE INDICATORS (From P. Diamond and R. Denman, *Laboratory Techniques in Chemistry and Biochemistry*, 2nd edn, Butterworths, London, (1972)

Indicator	pH range	Colour in acid solution	Colour in alkaline solution	pK_{In}	Working solution
Cresol red (acid)	0·2–1·8	Red	Yellow	—	0·1 g in 13·3 ml 0·02 M NaOH made up to 250 ml in water
Thymol blue (acid)	1·2–2·8	Red	Yellow	1·7	0·1 g in 10·75 ml 0·02 M NaOH made up to 250 ml in water
Bromophenol blue	2·8–4·6	Yellow	Blue	4·0	0·1 g in 7·5 ml 0·02 M NaOH made up to 250 ml in water
Methyl orange	3·1–4·4	Red	Yellow	3·7	0·1% in water
Congo red	3·0–5·0	Violet	Red	—	0·1% in water
Bromocresol green	3·8–5·4	Yellow	Blue	4·7	0·1 g in 7·25 ml 0·02 M NaOH made up to 250 ml in water
Methyl red	4·2–6·3	Red	Yellow	5·1	0·1 g in 18·6 ml 0·02 M NaOH made up to 250 ml in water
Bromocresol purple	5·2–6·8	Yellow	Purple	6·3	0·1 g in 9·25 ml 0·02 M NaOH made up to 250 ml in water
Bromothymol blue	6·0–7·6	Yellow	Blue	7·0	0·1 g in 8 ml 0·02 M NaOH made up to 250 ml in water
Phenol red	6·8–8·4	Yellow	Red	7·9	0·1 g in 14·3 ml 0·02 M NaOH made up to 250 ml in water
Cresol red (base)	7·2–8·8	Yellow	Red	8·3	As for acid indicator
Thymol blue (base)	8·0–9·6	Yellow	Blue	8·9	0·1 g in 10·75 ml 0·02 M NaOH made up to 250 ml in water
Phenolphthalein	8·3–10·0	Colourless	Red	9·6	0·1% in 50% aqueous ethanol
Thymolphthalein	8·3–10·5	Colourless	Blue	9·2	0·1% in 80% aqueous ethanol
Alizarine yellow R	10·1–12·0	Yellow	Red–orange	—	0·1% in water

APPENDIX VI

FACTORS FOR SOLUTIONS IN VOLUMETRIC ANALYSIS

Acids	Factor (g/ml of normal solution)
Acetic acid, CH_3COOH	0·06005
Benzoic acid, C_6H_5COOH	0·1221
Boric acid, H_3BO_3	0·06184
Citric acid, $C_6H_8O_7 \cdot H_2O$	0·07005
Hydrochloric acid, HCl	0·03646
Lactic acid, $C_3H_6O_3$	0·09008
Malic acid, $C_4H_6O_5$	0·06706
Oleic acid, $C_{18}H_{34}O_2$	0·28245
Phosphoric acid (ortho), H_3PO_4	0·04900
Sulphuric acid, H_2SO_4	0·04904
Tartaric acid, $C_4H_6O_6$	0·07504

Alkalis	Factor (g/ml of normal solution)
Ammonia, NH_3	0·01703 ($\equiv 0·014$ g N)
Barium hydroxide, $Ba(OH)_2$	0·08569
Calcium carbonate, $CaCO_3$	0·05004
Calcium oxide, CaO	0·02804
Potassium carbonate, K_2CO_3	0·06910
Potassium hydroxide, KOH	0·05610
Potassium oxide, K_2O	0·04710
Sodium bicarbonate, $NaHCO_3$	0·08401
Sodium carbonate, Na_2CO_3	0·05300
Sodium hydroxide, NaOH	0·04000

OTHER EQUIVALENTS

Calcium	1 ml 0·1 N permanganate $\equiv 0·002004$ g Ca
Calcium carbonate	1 ml 0·1 N permanganate $\equiv 0·005004$ g $CaCO_3$
Calcium oxide	1 ml 0·1 N permanganate $\equiv 0·002804$ g CaO

Chloride	1 ml 0·1 N silver nitrate	≡ 0·003546 g Cl
Iodine	1 ml 0·1 N silver nitrate	≡ 0·01269 g I
Ferrous iron	1 ml 0·1 N permanganate	≡ 0·005585 g Fe
Oxalic acid	1 ml 0·1 N permanganate	≡ 0·006303 g $C_2H_2O_4 \cdot 2H_2O$
Potassium chloride	1 ml 0·1 N silver nitrate	≡ 0·007456 g KCl
Sodium chloride	1 ml 0·1 N silver nitrate	≡ 0·005844 g NaCl
Sodium nitrite	1 ml 0·1 N permanganate	≡ 0·003450 g $NaNO_2$
Sodium sulphite	1 ml 0·1 N iodine	≡ 0·01261 g $Na_2SO_3 \cdot 7H_2O$
Sulphur dioxide	1 ml 0·1 N iodine	≡ 0·003203 g SO_2

NOTES ON
PREPARATION OF SOME STANDARD SOLUTIONS

Ammonia Solutions are prepared by diluting ammonia of sp. gr. 0·880, which contains 27–30% w/w of NH_3.

Bromine (0·5 N) 15 g of potassium bromate plus 75 g of potassium bromide per litre of water. For checking the strength, add KI and a slight excess of HCl and titrate the solution with 0·5 N sodium thiosulphate using starch as indicator near the end of the titration.

Hydrochloric acid Solutions are prepared by diluting concentrated HCl, which contains 35–38% w/w of HCl.

Iodine (0·1 N) Dissolve 18 g of KI in the minimum volume of water and add 12·69 g of iodine. Stir the mixture to dissolve the solid and dilute the solution with water to 1 litre.

Potassium bromate (0·1 N) Contains 2·7835 g of $KBrO_3$ per litre of water.

Potassium dichromate (0·1 N) Contains 4·903 g of $K_2Cr_2O_7$ per litre of water.

Sodium nitrite (0·1 M) Dissolve 6·900 g of $NaNO_2$ in cold water and dilute the solution to 1 litre.

Sodium thiosulphate (0·1 N) Contains 24·82 g of $Na_2S_2O_3 \cdot 5H_2O$ per litre of boiled and cooled distilled water. The solution is more stable if 4 g of borax and three drops of chloroform are added before diluting the solution to the mark.

APPENDIX VII

THE SI SYSTEM OF UNITS

The SI system is based on the following six primary units:

Quantity	Unit	Symbol
length	metre	m
mass	kilogram	kg
time	second	s
electric current	ampere	A
temperature	kelvin	K
luminous intensity	candela	cd

Decimal multiples and sub-multiples of the SI units are formed with the following prefixes:

Factor by which the unit is multiplied	Prefix	Symbol
10^{12}	tera	T
10^{9}	giga	G
10^{6}	mega	M
10^{3}	kilo	k
10^{2}	hecto	h
10	deca	da
10^{-1}	deci	d
10^{-2}	centi	c
10^{-3}	milli	m
10^{-6}	micro	μ
10^{-9}	nano	n
10^{-12}	pico	p
10^{-15}	femto	f
10^{-18}	atto	a

CONVERSION OF COMMON UNITS TO EQUIVALENTS IN SI UNITS

LENGTH

1 ft	= 0·3048 m
1 in	= 25·4 mm = 2·54 cm

AREA

1 ft^2	= 0·09290 m^2 = 929·0 cm^2
1 in^2	= 645·2 mm^2 = 6·452 cm^2

VOLUME

1 ft^3	= 0·028317 m^3 = 28·317 dm^3
1 in^3	= 16·387 cm^3

MASS

1 ton	= 1016·05 kg
1 cwt	= 50·80 kg
1 lb	= 0·45359 kg = 453·6 g
1 oz	= 28·35 g
1 dr (dram)	= 1·772 g
1 gr (grain)	= 64·80 mg = 0·0648 g
1 oz apoth = 1 oz tr	= 31·10 g
1 drachm	= 3·888 g

CAPACITY

1 gallon	= 4·546 dm^3 or litres
1 pint	= 0·5683 dm^3 or litre = 568·3 cm^3
1 fluid ounce	= 28·41 cm^3
1 fluid drachm	= 3551·6 mm^3 = 3·5516 cm^3
1 minim	= 59·194 mm^3

TEMPERATURE

$$°C: \quad \theta = \frac{5}{9}(t-32)$$

$$K: \quad T = \frac{5}{9}(t+459{\cdot}67)$$

$$°R: \quad r = t+459{\cdot}67$$

t = Temperature on Fahrenheit scale
r = Temperature on Rankine scale (°R), i.e., absolute Fahrenheit
T = Temperature on Kelvin scale
θ = Temperature on Celsius scale (°C)

The zero on the Celsius scale is the ice-point (273·15 K)

ENERGY

The joule (J) is the unit of energy in the SI system
1 cal (15 °C calorie) = 4·186 J
1 kcal = 4·186 kJ

ATOMIC WEIGHTS OF THE ELEMENTS

Atomic no.	Element	Symbol	Atomic weight	Atomic no.	Element	Symbol	Atomic weight
1	Hydrogen	H	1·00797	52	Tellurium	Te	127·60
2	Helium	He	4·0026	53	Iodine	I	126·9044
3	Lithium	Li	6·939	54	Xenon	Xe	131·30
4	Beryllium	Be	9·0122	55	Caesium	Cs	132·905
5	Boron	B	10·811	56	Barium	Ba	137·34
6	Carbon	C	12·01115	57	Lanthanum	La	138·91
7	Nitrogen	N	14·0067	58	Cerium	Ce	140·12
8	Oxygen	O	15·9994	59	Praseodymium	Pr	140·907
9	Fluorine	F	18·9984	60	Neodymium	Nd	144·24
10	Neon	Ne	20·183	61	Promethium	Pm	(147)
11	Sodium	Na	22·9898	62	Samarium	Sm	150·35
12	Magnesium	Mg	24·312	63	Europium	Eu	151·96
13	Aluminium	Al	26·9815	64	Gadolinium	Gd	157·25
14	Silicon	Si	28·086	65	Terbium	Tb	158·924
15	Phosphorus	P	30·9738	66	Dysprosium	Dy	162·50
16	Sulphur	S	32·064	67	Holmium	Ho	164·930
17	Chlorine	Cl	35·453	68	Erbium	Er	167·26
18	Argon	Ar	39·948	69	Thulium	Tm	168·934
19	Potassium	K	39·102	70	Ytterbium	Yb	173·04
20	Calcium	Ca	40·08	71	Lutetium	Lu	174·97
21	Scandium	Sc	44·956	72	Hafnium	Hf	178·49
22	Titanium	Ti	47·90	73	Tantalum	Ta	180·948
23	Vanadium	V	50·942	74	Tungsten	W	183·85
24	Chromium	Cr	51·996	75	Rhenium	Re	186·2
25	Manganese	Mn	54·9380	76	Osmium	Os	190·2
26	Iron	Fe	55·847	77	Iridium	Ir	192·2
27	Cobalt	Co	58·9332	78	Platinum	Pt	195·09
28	Nickel	Ni	58·71	79	Gold	Au	196·967
29	Copper	Cu	63·54	80	Mercury	Hg	200·59
30	Zinc	Zn	65·37	81	Thallium	Tl	204·37
31	Gallium	Ga	69·72	82	Lead	Pb	207·19
32	Germanium	Ge	72·59	83	Bismuth	Bi	208·980
33	Arsenic	As	74·9216	84	Polonium	Po	(210)
34	Selenium	Se	78·96	85	Astatine	At	(210)
35	Bromine	Br	79·909	86	Radon	Rn	(222)
36	Krypton	Kr	83·80	87	Francium	Fr	(223)
37	Rubidium	Rb	85·47	88	Radium	Ra	(226)
38	Strontium	Sr	87·62	89	Actinium	Ac	(227)
39	Yttrium	Y	88·905	90	Thorium	Th	232·038
40	Zirconium	Zr	91·22	91	Protactinium	Pa	(231)
41	Niobium	Nb	92·906	92	Uranium	U	238·03
42	Molybdenum	Mo	95·94	93	Neptunium	Np	(237)
43	Technetium	Tc	(99)	94	Plutonium	Pu	(242)
44	Ruthenium	Ru	101·07	95	Americium	Am	(243)
45	Rhodium	Rh	102·905	96	Curium	Cm	(247)
46	Palladium	Pd	106·4	97	Berkelium	Bk	(247)
47	Silver	Ag	107·870	98	Californium	Cf	(249)
48	Cadmium	Cd	112·40	99	Einsteinium	Es	(254)
49	Indium	In	114·82	100	Fermium	Fm	(253)
50	Tin	Sn	118·69	101	Mendelevium	Md	(256)
51	Antimony	Sb	121·75	102	Nobelium	No	(253)
				103	Lawrencium	Lw	(257)

The figures given in parentheses are the mass numbers of the most stable or most common isotope.

	0	1	2	3	4	5	6	7	8	9	1	2	3	4	5	6	7	8	9
10	0000	0043	0086	0128	0170	0212	0253	0294	0334	0374	4	8	12	17	21	25	29	33	37
11	0414	0453	0492	0531	0569	0607	0645	0682	0719	0755	4	8	11	15	19	23	26	30	34
12	0792	0828	0864	0899	0934	0969	1004	1038	1072	1106	3	7	10	14	17	21	24	28	31
13	1139	1173	1206	1239	1271	1303	1335	1367	1399	1430	3	6	10	13	16	19	23	26	29
14	1461	1492	1523	1553	1584	1614	1644	1673	1703	1732	3	6	9	12	15	18	21	24	27
15	1761	1790	1818	1847	1875	1903	1931	1959	1987	2014	3	6	8	11	14	17	20	22	25
16	2041	2068	2095	2122	2148	2175	2201	2227	2253	2279	3	5	8	11	13	16	18	21	24
17	2304	2330	2355	2380	2405	2430	2455	2480	2504	2529	2	5	7	10	12	15	17	20	22
18	2553	2577	2601	2625	2648	2672	2695	2718	2742	2765	2	5	7	9	12	14	16	19	21
19	2788	2810	2833	2856	2878	2900	2923	2945	2967	2989	2	4	7	9	11	13	16	18	20
20	3010	3032	3054	3075	3096	3118	3139	3160	3181	3201	2	4	6	8	11	13	15	17	19
21	3222	3243	3263	3284	3304	3324	3345	3365	3385	3404	2	4	6	8	10	12	14	16	18
22	3424	3444	3464	3483	3502	3522	3541	3560	3579	3598	2	4	6	8	10	12	14	15	17
23	3617	3636	3655	3674	3692	3711	3729	3747	3766	3784	2	4	6	7	9	11	13	15	17
24	3802	3820	3838	3856	3874	3892	3909	3927	3945	3962	2	4	5	7	9	11	12	14	16
25	3979	3997	4014	4031	4048	4065	4082	4099	4116	4133	2	3	5	7	9	10	12	14	15
26	4150	4166	4183	4200	4216	4232	4249	4265	4281	4298	2	3	5	7	8	10	11	13	15
27	4314	4330	4346	4362	4378	4393	4409	4425	4440	4456	2	3	5	6	8	9	11	13	14
28	4472	4487	4502	4518	4533	4548	4564	4579	4594	4609	2	3	5	6	8	9	11	12	14
29	4624	4639	4654	4669	4683	4698	4713	4728	4742	4757	1	3	4	6	7	9	10	12	13
30	4771	4786	4800	4814	4829	4843	4857	4871	4886	4900	1	3	4	6	7	9	10	11	13
31	4914	4928	4942	4955	4969	4983	4997	5011	5024	5038	1	3	4	6	7	8	10	11	12
32	5051	5065	5079	5092	5105	5119	5132	5145	5159	5172	1	3	4	5	7	8	9	11	12
33	5185	5198	5211	5224	5237	5250	5263	5276	5289	5302	1	3	4	5	7	8	9	10	12
34	5315	5328	5340	5353	5366	5378	5391	5403	5416	5428	1	3	4	5	6	8	9	10	11
35	5441	5453	5465	5478	5490	5502	5514	5527	5539	5551	1	2	4	5	6	7	9	10	11
36	5563	5575	5587	5599	5611	5623	5635	5647	5658	5670	1	2	4	5	6	7	8	10	11
37	5682	5694	5705	5717	5729	5740	5752	5763	5775	5786	1	2	3	5	6	7	8	9	10

305

N	0	1	2	3	4	5	6	7	8	9	1	2	3	4	5	6	7	8	9
38	5798	5809	5821	5832	5843	5855	5866	5877	5888	5899	1	2	3	5	6	7	8	9	10
39	5911	5922	5933	5944	5955	5966	5977	5988	5999	6010	1	2	3	4	5	7	8	9	10
40	6021	6031	6042	6053	6064	6075	6085	6096	6107	6117	1	2	3	4	5	6	8	9	10
41	6128	6138	6149	6160	6170	6180	6191	6201	6212	6222	1	2	3	4	5	6	7	8	9
42	6232	6243	6253	6263	6274	6284	6294	6304	6314	6325	1	2	3	4	5	6	7	8	9
43	6335	6345	6355	6365	6375	6385	6395	6405	6415	6425	1	2	3	4	5	6	7	8	9
44	6435	6444	6454	6464	6474	6484	6493	6503	6513	6522	1	2	3	4	5	6	7	8	9
45	6532	6542	6551	6561	6571	6580	6590	6599	6609	6618	1	2	3	4	5	6	7	8	9
46	6628	6637	6646	6656	6665	6675	6684	6693	6702	6712	1	2	3	4	5	6	6	7	8
47	6721	6730	6739	6749	6758	6767	6776	6785	6794	6803	1	2	3	4	5	5	6	7	8
48	6812	6821	6830	6839	6848	6857	6866	6875	6884	6893	1	2	3	4	4	5	6	7	8
49	6902	6911	6920	6928	6937	6946	6955	6964	6972	6981	1	2	3	4	4	5	6	7	8
50	6990	6998	7007	7016	7024	7033	7042	7050	7059	7067	1	2	3	3	4	5	6	7	8
51	7076	7084	7093	7101	7110	7118	7126	7135	7143	7152	1	2	3	3	4	5	6	7	8
52	7160	7168	7177	7185	7193	7202	7210	7218	7226	7235	1	2	2	3	4	5	6	6	7
53	7243	7251	7259	7267	7275	7284	7292	7300	7308	7316	1	2	2	3	4	5	6	6	7
54	7324	7332	7340	7348	7356	7364	7372	7380	7388	7396	1	2	2	3	4	5	6	6	7
55	7404	7412	7419	7427	7435	7443	7451	7459	7466	7474	1	2	2	3	4	5	5	6	7
56	7482	7490	7497	7505	7513	7520	7528	7536	7543	7551	1	2	2	3	4	5	5	6	7
57	7559	7566	7574	7582	7589	7597	7604	7612	7619	7627	1	2	2	3	4	5	5	6	7
58	7634	7642	7649	7657	7664	7672	7679	7686	7694	7701	1	1	2	3	4	4	5	6	7
59	7709	7716	7723	7731	7738	7745	7752	7760	7767	7774	1	1	2	3	4	4	5	6	7
60	7782	7789	7796	7803	7810	7818	7825	7832	7839	7846	1	1	2	3	4	4	5	6	6
61	7853	7860	7868	7875	7882	7889	7896	7903	7910	7917	1	1	2	3	4	4	5	6	6
62	7924	7931	7938	7945	7952	7959	7966	7973	7980	7987	1	1	2	3	3	4	5	6	6
63	7993	8000	8007	8014	8021	8028	8035	8041	8048	8055	1	1	2	3	3	4	5	5	6
64	8062	8069	8075	8082	8089	8096	8102	8109	8116	8122	1	1	2	3	3	4	5	5	6
65	8129	8136	8142	8149	8156	8162	8169	8176	8182	8189	1	1	2	3	3	4	5	5	6
66	8195	8202	8209	8215	8222	8228	8235	8241	8248	8254	1	1	2	3	3	4	5	5	6
67	8261	8267	8274	8280	8287	8293	8299	8306	8312	8319	1	1	2	3	3	4	5	5	6

Proportional parts (mean differences):

```
6 6 | 6 5 5 5 5 | 5 5 5 5 5 | 5 5 5 5 5 | 5 5 5 4 4 | 4 4 4 4 4 | 4 4 4 4 4
5 5 | 5 5 5 5 5 | 5 5 4 4 4 | 4 4 4 4 4 | 4 4 4 4 4 | 4 4 4 4 3 | 3 3 3 3 3
4 4 | 4 4 4 4 4 | 4 4 4 4 4 | 4 4 4 4 4 | 4 4 3 3 3 | 3 3 3 3 3 | 3 3 3 3 3
------------------------------------------------------------------------------
4 4 | 4 4 4 4 4 | 3 3 3 3 3 | 3 3 3 3 3 | 3 3 3 3 3 | 3 3 3 3 3 | 3 3 3 3 3
3 3 | 3 3 3 3 3 | 3 3 3 3 3 | 3 3 3 3 3 | 3 3 2 2 2 | 2 2 2 2 2 | 2 2 2 2 2
3 2 | 2 2 2 2 2 | 2 2 2 2 2 | 2 2 2 2 2 | 2 2 2 2 2 | 2 2 2 2 2 | 2 2 2 2 2
------------------------------------------------------------------------------
2 2 | 2 2 2 2 2 | 2 2 2 2 2 | 2 2 2 2 2 | 2 2 1 1 1 | 1 1 1 1 1 | 1 1 1 1 1
1 1 | 1 1 1 1 1 | 1 1 1 1 1 | 1 1 1 1 1 | 1 1 1 1 1 | 1 1 1 1 1 | 1 1 1 1 1
1 1 | 1 1 1 1 1 | 1 1 1 1 1 | 1 1 0 0 0 | 0 0 0 0 0 | 0 0 0 0 0 | 0 0 0 0 0
```

N	0	1	2	3	4	5	6	7	8	9
68	8325	8331	8338	8344	8351	8357	8363	8370	8376	8382
69	8388	8395	8401	8407	8414	8420	8426	8432	8439	8445
70	8451	8457	8463	8470	8476	8482	8488	8494	8500	8506
71	8513	8519	8525	8531	8537	8543	8549	8555	8561	8567
72	8573	8579	8585	8591	8597	8603	8609	8615	8621	8627
73	8633	8639	8645	8651	8657	8663	8669	8675	8681	8686
74	8692	8698	8704	8710	8716	8722	8727	8733	8739	8745
75	8751	8756	8762	8768	8774	8779	8785	8791	8797	8802
76	8808	8814	8820	8825	8831	8837	8842	8848	8854	8859
77	8865	8871	8876	8882	8887	8893	8899	8904	8910	8915
78	8921	8927	8932	8938	8943	8949	8954	8960	8965	8971
79	8976	8982	8987	8993	8998	9004	9009	9015	9020	9025
80	9031	9036	9042	9047	9053	9058	9063	9069	9074	9079
81	9085	9090	9096	9101	9106	9112	9117	9122	9128	9133
82	9138	9143	9149	9154	9159	9165	9170	9175	9180	9186
83	9191	9196	9201	9206	9212	9217	9222	9227	9232	9238
84	9243	9248	9253	9258	9263	9269	9274	9279	9284	9289
85	9294	9299	9304	9309	9315	9320	9325	9330	9335	9340
86	9345	9350	9355	9360	9365	9370	9375	9380	9385	9390
87	9395	9400	9405	9410	9415	9420	9425	9430	9435	9440
88	9445	9450	9455	9460	9465	9469	9474	9479	9484	9489
89	9494	9499	9504	9509	9513	9518	9523	9528	9533	9538
90	9542	9547	9552	9557	9562	9566	9571	9576	9581	9586
91	9590	9595	9600	9605	9609	9614	9619	9624	9628	9633
92	9638	9643	9647	9652	9657	9661	9666	9671	9675	9680
93	9685	9689	9694	9699	9703	9708	9713	9717	9722	9727
94	9731	9736	9741	9745	9750	9754	9759	9763	9768	9773
95	9777	9782	9786	9791	9795	9800	9805	9809	9814	9818
96	9823	9827	9832	9836	9841	9845	9850	9854	9859	9863
97	9868	9872	9877	9881	9886	9890	9894	9899	9903	9908
98	9912	9917	9921	9926	9930	9934	9939	9943	9948	9952
99	9956	9961	9965	9969	9974	9978	9983	9987	9991	9996

INDEX

Abbreviations, xi
Acetic acid in vinegar, 257
Acid-insoluble ash, 58
Acid phosphates, 228–230
Acid value of oils and fats, 125
Acidity, 70
  of cheese, 160
  of milk, 141
  of vinegar, 257
Acids in jam, 245
Action limit, 22–24
Additives, 78–96
Adulteration of fruit juices, 268
Agricultural control, 2, 4
Agtron, 220
AIS, 274
Alcohol, determination of, 74
Alcohol-insoluble solids, 274
Alkalinity of the soluble ash, 57
Alkaloids in cocoa, 276
Alumina cream, 61
Aneurine in flour, 227
Animal charcoal as clearing agent, 61
Antioxidants, 87, 286
AQL, 22
Arsenic, 99, 286
Arterial pumping, 195
Artificial sweeteners, 288
Ascorbic acid, 267
Ash, determination of, 56
Atomic weights of the elements, 303
AutoAnalyser, 10–12
Average control chart, 23–24

Bailey-Walker method for fat, 44
Baking powder, 228, 287
  ACP in, 229

Baking powder *continued*
  ASP in, 229
  filler in, 228
  raw materials, 228
  sodium bicarbonate in, 228
Barley as filler, 176
Beef, 173, 176
Beef sausage, meat in, 176
Beer's law, 291
Benzoic acid, 84
Blish and Sandstedt method, 223
Bloater, 207
Bloom gelometer, 283
Bolton method for fat, 44, 178
Borax buffer, 72
Boucher jelly tester, 282
Bread, 287
Brineometer, 197
Brining, 194
Brix hydrometer, 252–255
Buffer salts in jam, 245
Bulk density, of dried milk, 151
  of sugar, 240
Butter, 152, 287
  salt in, 153
Butter fat, semi-micro Reichert–Polenske
  processes, 153
Butylated hydroxyanisole, 88
Butylated hydroxytoluene, 89
Butyrometers, Gerber, 133, 135, 149,
  150, 153, 157, 161
Butyrometers, van Gulik, 45, 179
Buying sample, 5

Caffeine, in coffee products, 274
  in tea, 276
Calcium, 102

Calcium *continued*
  in flour, 225
  in salt, 185, 196
Calorie, 294, 302
Canned fruits, 251
Canned meats, 176, 189
Carbohydrate, 188, 294
Carbon dioxide, in baking powder, 230–234
  in self-raising flour, 234
Carbon dioxide, density of, 232
Carcinogenic substances, 295
Carrez solution, 60
Carter-Simon oven, 33
Casings for sausages, 186
Cereal filler in flesh products, 175, 185
Chalk, in flour, 225
  in self-raising flour, 235
Cheese, 156, 287
  acidity of, 160
  fat in, 157
  salt in, 159
Chili saltpetre, 197
Chittick apparatus, 230
Chlorine, free, in water, 272
Chocolate flour confectionery, 277
Citrus fruit juices, 262
Clarification of solutions for sugar analysis, 60
Clearing agents, 60
Clerget–Herzfeld method, 69
Cocoa content, assessment of, 276
Codes of practice, 289
Coffee, 274
Coffee and chicory mixture, 276, 287
Coffee essences, 274, 287
Coffee extracts, 274
Coffee, instant, 275
Coffee products, caffeine in, 274
Colour, of oils, 119
  of sugar, 242
Colouring matters, 90
  extraction from foods, 95
  paper chromatography of, 90
  spectrophotometry, 94
  thin-layer chromatography of, 93
Colours, 90, 287
  in jam, 245
Competitor's samples, 16
Complaint samples, 16
Condensed milk, 148, 287
  fat in, 148
Consistency, 2

Contract crop, 4
Control, agricultural, 4
Control, process, 9
Control, storage, 15
Conway microdiffusion technique, 55, 170
Copper, 103
Cream, 146, 287
  fat in, 146–147
Cream of tartar in baking powder, 228
Cream powders, 228
Creatine and creatinine in meat extract, 280
Crude fibre, 48
Crude protein, 50
Cured meat, 201
Curry powder, 287
Cut-out in canned fruits, 254

Defatted meat content, 188
Dextrose, 65, 68
Diastatic activity of flour, 223
Diphenyl, 86
Diphenylcarbazone, indicator, 199
Distilled vinegar, 257
Dodecyl gallate, 88
Dogfish, 174
Drained weight of canned fruits, 255
Dried milk, 149, 287
  bulk density of, 151
  fat in, 150
  solubility of, 150
Dry ashing for destruction of organic matter, 98
Dry curing, 196
Dye-binding method for protein, 56, 182
Dyes, 90

Egg content, assessment of, 278
Egg, dried, solubility of, 280
Egg, liquid, 287
Egg yolk solids, 279
Elasmobranch fish, 174
Emulsifiers, 287
Emulsion method of curing, 196
Energy calculations, 294
Enzymic activity of fruit juices, 263
Equilibrium relative humidity (ERH), 41
Equivalent pints of condensed milk, 148
Ester value of vinegar, 257
Ethanol, determination of, 74–77
Ether extract, determination of, 43, 45

Evaporated milk, 148
  fat in, 148
Extensograph, 7
Extinction coefficient, 291
Extraneous matter, 13

Factors for volumetric analysis, 298
Farinograph, 7
Fat, determination of, 43
  (see also under individual foods)
Fat spoilage values from chloroform extract, 127
Fehling's solution, 62
  standardisation of, 64
Fibre, 48, 215
Filled milks, 288
Filled weights, 255
Fillers, 175, 176, 190
Filters for absorptiometry, 290
Fish, 166
  composition of, 174
  spoilage, 167
Fish cakes, 192, 287
Fish content of products, 175
Fish curing, 206
Fish paste, 191, 287
Fish products, analysis of, 175, 191–194, 206
Fish smoking, 207
Fixed acidity, 71, 262
Fixed ether extract, 45
Flesh foods, 166–212
  sampling of, 167
  spoilage of, 169
Flour, 213, 287
  acidity of, 218
  ash of, 215
  assessment of grade, 215, 220
  bleach value, 222
  chalk in, 225
  conditioning, 215
  diastatic activity, 223
  fibre in, 215
  fortification of, 214, 226
  gluten in, 216
  maltose value of, 223
  milling, 214
  moisture of, 215
  particle size, 222
  Pekar test, 219
  pH of, 218, 219
  protein, 216

Flour continued
  self-raising, 234, 287
  thiamine in 227
Fluorine, 97, 287
Food and Drugs Act, 286
Food additives, 78
Food Additives and Contaminants Committee, 289
Food energy, 294
Food legislation, 286
Food Standards Committee, 289
Formol titration, 56, 162, 257
Free fatty acids (FFA), 125, 172
Freezing point of milk, 139
Freshness of oils, 124
Freshwater fish, 169
Fruit curd, 287
  egg in, 278
Fruit juices, 262
Fruits, canned, 251

Gallates, 88
Gaussian distribution, 22
Gelatine, 287
  jelly strength of, 281
Gerber method, 45
  for cheese, 157
  for condensed milk, 148
  for cream, 147
  for milk powder, 150
  for milk, 132
Gluten in flour, 216
Golden raising powder, 230, 287
Grapefruit juice, 264
Griess–Ilosvay method for nitrites, 199

Haenni method for dried egg, 280
Hagberg falling number test, 224
Hardness of water, 270
Hartley funnel, 49, 278
Headspace in canned fruits, 252
Herring, 175
Homogenised milk (fat estimation), 134
Hortvet method, 139
Howard mould count, 258
Hydrogen swells, 252
Hydrometer, Brix (for sugar), 252
Hydrometer scales for salt, 198
Hydroxymethylfurfural in fruit juices, 265
Hygiene, food, 287

Ice-cream, 160, 287
  fat in, 161
  formol titration for, 162
  milk solids in, 162
  overrun of, 162
ICUMSA method for the colour of
  sugar, 242
Ilford spectrum filters, 291
Indicators, 296
Insoluble matter in sugar, 239
Insoluble solids in fruits, 250
Instant coffee, 275, 287
Intermediates, 9
Inversion of sucrose, 64, 69
Invert sugar, 68
  in jam, 249
  in sugar, 237
Iodine value of vinegar, 257
Iodine values of oils and fats, 120
Iron, 105
  in flour, 226
Irradiation, 287

Jam, 242, 287
  fruit in, 243, 250
  ingredients in, 243
  insoluble solids in, 251
  pectin in, 244
  pH of, 245
  reducing sugars in, 249
  soluble solids of, 248
  sugar in, 243
Jelly strength of gelatine, 281
Joule, 294, 302

Karl Fischer method, 37
Kent–Jones and Martin flour colour
  grader, 220
Kippers, 207
Kirschner value, 156
Kjeldahl method, 50
  macro procedure, 52
  semi-micro procedure, 54
Kreis test, 127

Labelling, 287
Lactic acid in milk, 141
Lactometer, 133
Lactose, 65, 68
LAJAC Code of Practice for canned
  fruits, 252

Lambert's law, 291
Lane and Eynon's process, 62
  calculation for, 67
  tables of factors for, 65, 66
Lea method for peroxide value, 125
Lead, 97, 106, 287
Lead acetate clearing agents, 61
Lean meat content, 177
Legislation, 286
Lemon curd, 278, 287
Lemon juice, 264
Lower action limit, 24
Lower warning limit, 24
Lucke and Geidel method, 169
Luncheon meat, 176

Macdonald method for fat, 45
  in ice cream, 161
Magnesium in salt, 185, 196
Malt vinegar, 257, 258
Maltose value of flour, 223
Margarine, 156, 287
Markham apparatus, 54
Marmalade, 243, 287
Maturity ratio, 263
Mayonnaise, 288
Meade's method for bulk density of
  sugar, 241
Meat, 166
  canned, 288
  composition of, 173
  sampling of, 167, 177
  spoilage, 167
  water in, 174, 177
Meat brines, 197
Meat content of products, 176
Meat extract, creatine in, 280
Meat paste, meat in, 191, 288
Meat pie, meat in, 190, 288
Meat products, analysis of, 177
Meat treatment, 288
Meats, cured, 201
Mercuric nitrate method for salt, 198
Mercury, 207
Milk, 131, 288
  acidity of, 141
  added water in, 140
  Channel Island, 288
  condensed, 148, 287
  density, 132, 136
  dried, 149, 287
  evaporated, 148

Milk *continued*
    fat in, 132–136
    filled, 288
    freezing point of, 139
    methylene blue test, 142, 143
    pasteurised, 131, 143, 288
    phosphatase test, 143
    powder, 149
    resazurin test, 142
    sampling, 132
    solids-not-fat of, 136
    South Devon, 288
    standards, 137, 288
    sterilised, 144–146, 288
    ultra-heat treated (UHT), 131, 144, 288
Milk powder, 149
Milk solids in ice-cream, 162
Mincemeat, 288
Mineral matter (ash), 56
Mineral oil, 124, 288
Moisture, determination of, 28–41
Mojonnier method for fat, 134
Mojonnier tube, 87
Monier–Williams method for sulphur dioxide, 79
Mould in jam, 249
Mustard, 288
Mutarotation, 69

Neutralising value of acid phosphates, 229
Nicotinic acid in flour, 214, 228
Nitrate, 197
    in brines, 196
    in cured meats, 205
Nitrite, 197
    in brines, 197
    in cured meats, 202
Nitrogen, determination of, 50
    in fillers, 175, 176

Octyl gallate, 88
Oil-soluble colours, 96
Oil values, 119
Oils, 121, 122
Optical density, 292
Optical rotation, 68
Orange juice, 264
Organic phosphorus in egg, 279

Orthophenylphenol, 86
Overrun of ice-cream, 162
Oxidation – reduction potential, 73
Oxidation value of vinegar, 257
Oxidative rancidity of oils, 126

Palin's method for free chlorine, 273
Parahydroxybenzoates, 86
Pasteurised milk, 131, 143, 288
    phosphatase test for, 143
Pectin in jam, 242, 244
Pectolytic activity, 263
Pekar colour test for flour, 219
Peroxide value of oils, 125, 128
Pesticides, 284
Pfeifer-Wartha method for hardness of water, 271
pH value, 72
    of brines, 201
    of flour, 218, 219
    of fruits, 250
    of jam, 249
Phosphatase test for pasteurised milk, 143
Phosphate, 108–111
Phosphorus, 108
    organic, in egg, 278
Phosphotungstic acid clearing agent, 61
Phthalate buffer, 72
Pickles, 256
    acidity of, 261
    salt in, 261
    total volatiles of, 261
Polarimetry of sugars, 68
Pork, 173
Potassium, 111
Potassium nitrite, 197
Potato filler, 176, 190
Preservatives, 78, 288
Process control, 9
Product control, 14
Proof spirit, 74
Propionic acid, 86
Propyl gallate, 88
Protein, determination of, 50
Public analysts, 289

Quality, 2
Quality control, 3
    history of, 3

Quality control charts, 22
Quantabs, 10
Quartering, 27, 28

Rancidity of oils and fats, 124
Range control chart, 23
Raw materials, examination of, 5
Recknagel's phenomenon, 132
Recording of samples, 18
Redox potential, 73
    of brines, 201
Reduced iron, 226
Reducing sugars, 61–70
    in jam, 249
    in sugar, 237
Refractive index, of fruit juices, 263, 264
    of sugar solutions, 58
Reporting, 18
Resazurin test for milk, 142
Residual carbon dioxide, in baking powder, 233
    in self-raising flour, 234
R_F values of colours, 91, 93, 95
Richmond's equation for milk, 136
Ripper's titration for sulphur dioxide, 84
Rose–Gottlieb method, 44
    for milk, 134
Rusk filler in flesh products, 176, 185

Saccharin tablets, 288
Salad cream, 288
    egg in, 278
Sale of Goods Act, 5
Salinometer, 197, 198
Salmon, 175
Salt, analysis, 185
    in cured meats, 201
    in meat curing, 196
Saltpetre, 196
Sample preparation, 27
Sampling, 27
    (see also individual foods)
Sandiness in condensed milk, 148
Saponification value, determination of, 122
Saponification values of oils and fats, 122
Sauces, 256
Sausage rolls, 190, 288
    meat in, 190
Sausages, 184, 288
    analysis of, 186

Self-raising flour, 234, 287
Shipton's method for sulphur dioxide, 79
SI system of units, 300
Simon automatic gluten washer, 217
Smoking of fish, 207
Sodium bicarbonate, 229
Sodium chloride, 185
Sodium nitrate, 197
Sodium nitrite, 197
Sodium tungstate as clearing agent, 61
Soft drinks, 288
Solids-not-fat in milk, 136
Solubility, of dried egg, 279
    of milk powder, 150
Soluble solids, 58
    in jam, 248
Solvent removal (diagram), 135
Solvents, 288
Sorbic acid, 86
South Devon milk, 288
Soxhlet method, 44, 178
Soya in sausages, 188
Specific gravity of fruit juices, 264
Specific rotation of sugars, 68
Spectrophotometry, 290
Spices, 186
Spirits, 74
Spoilage, of fish, 167
    of meat, 167
Springers, 252
Squashes, 265, 268
Stabilisers, 287
Stack-burning 252
Stainless steel, 8
Standard solutions, preparation of, 298
Starch in flesh products, 183
Sterilised milk, 144
    reducing capacity of, 148
    turbidity test for, 144
Storage controls, 15
Sucrose, determination of, 67
    by double polarisation, 69
    by refractometry, 58
    by volumetric methods, 67
Suet, 288
Sugar, examination of, 237
Sugar syrups, boiling points of, 253
    Brix scales, 253
    hydrometer scales, 253
Sugars, determination of, 58
    by polarimetry, 68
    by refractometry, 58
    by volumetric methods, 61

Sully method for peroxide value, 129
Sulphated ash, 58
Sulphite preservative, 185
Sulphur dioxide, 78
  distillation of, 79, 81
  in fruit juices, 268
  in fruit pulps, 243–244
  in jam, 250
  in sausages, 185
  in sugar, 239
  titration of, 84, 187, 268
Sweeteners, artificial, 288
Syrup in canned fruits, 252–256
Syrups, 252–254

Tartaric acid in baking powder, 228
TBA number, 172
Tea, caffeine in, 276
Technicon AutoAnalyzer, 10
Thiamine in flour, 227
Thiobarbituric acid number, 172
Tin, 113
Tinctorial power of colours, 245–247
Tomato ketchup, 262, 288
Tomato purée, 258
  acidity of, 258
  colour of, 260
  salt in, 258
  Sipple's tables for, 259
  total solids in, 258–259
Total alkaloids in cocoa, 276
Total ash, determination of, 56
Total carbon dioxide, in baking powder, 230
  in self-raising flour, 234, 235
Total titratable acidity, 70
Total volatile bases (nitrogen), 169–172
Total volatiles of pickles, 261
Trace elements, 97
Trade Descriptions Act, 289
Transmission, 292
Trial batches, 8
Trimethylamine in fish, 168, 170, 172
Tuna, mercury in, 207
Turbidity test for sterilised milk, 144

Ultra heat-treated (UHT) milk, 131, 144, 288
Units, 300
Unsaponifiable matter, 123
Upper action limit, 24
Upper warning limit, 24
Urea in fish, 168

Van Gulik butyrometer, 179–181
Vegetable, canned, alcohol-insoluble solids (AIS) in, 274
Vinegar, 257
  alkaline oxidation value of, 257
  formol titration in, 257
  total acidity of, 257
Viscosity, 9
Vitamin B_1 in flour, 214, 226
Vitamin C in fruit juices, 266
Volatile acidity, 71, 262
  in aqueous phase, 72, 262
Volatile oil, 45
Volumetric analysis, factors for, 298

Warning limit, 24
Water, estimation of, 28
  free, 30
  free chlorine in, 272
  hardness of, 270
  in foods, 29
Water soluble ash, 57
Weights and Measures Act, 289
Werner–Schmid method, 44
  for cheese, 159
Wet immersion brining, 194
Wet oxidation of foods, 98, 209
Wheat flour, 213–228
Wheat grain, structure of, 213
White fish, 167–169
Wijs reagent, 120
Wrappers, 15

Zinc, 116
Zinc ferrocyanide clearing agent, 60